THE LONGEST FLIGHT

YUMA'S QUEST FOR THE FUTURE

SHIRLEY WOODHOUSE MURDOCK

JAMES A. GILLASPIE

For ordering information, contact:

Shirley W. Murdock

P.O. Box 7095

Roll, AZ 85347

Phone: 520/785-9531

Fax: 520/785-9540

Email: shirleym31@aol.com

— OR —

James A. Gillaspie

3030 Arizona Ave.

Yuma, AZ 85364

Phone: 520/726-5599

Fax: 520/726-1628

To the memory of

Paul Burch

He was many things to many people,

Husband, father, friend, mentor, teacher, gentleman

And one darn good airplane mechanic who knew no bounds.

The success of the endurance flight was his like no other.

About the Authors
Shirley Woodhouse Murdock

Shirley Woodhouse Murdock grew up on the same farm that she lives on now with her husband, Hubert ("George") Murdock. She attended the Mohawk Valley School, which is next to their farm in Roll, and commuted the fifty miles to Yuma Union High School on a school bus. She graduated from Colorado Woman's College in Denver, married a year after that and moved back to the farm in 1954 when her parents, Harold and Ethelind Woodhouse, were killed in their own Cessna. George and Shirley raised two sons, Kenny and Jimmy, and two daughters, Carol Fisher, and Janet Taylor, all of whom have families of their own now, totaling a dozen children.

Shirley grew up in a flying family. Her father and mother were interested in aviation for several decades before getting their private pilot's licenses in 1944 and 1949 respectively. Her brother, Bob, got his Navy Wings in 1945. When Shirley's mother was taking flying lessons from Betty Tank, their friend, George Murdock, was doing the same thing and he got his private pilot's license that same year. Shirley took a few lessons, but got busy with college and didn't get her license. She loved flying with her family, and still does.

Her sons both fly, Kenny having earned his private pilot's license at age 21. Kenny and Jimmy have a Super Cub and the family has owned an S-35, "V-tail" Bonanza for about 25 years. Both sons are partners with George in the family farming operation, Murdock Farms.

Shirley's main hobby, for a number of years, has been writing a larger book, about the history of Roll and the Mohawk Valley, including stories of her parents' pioneering there in the 1920's. Her working title is "On a Roll in the Mohawk Valley." The book is gradually moving toward publication.

About the Authors
Jim Gillaspie

Jim Gillaspie was born in Tularosa, NM, and went to grammar school and high school in three different states. He was attending Yuma Union High School when the endurance flight took place; however his family moved to Socorro, New Mexico a couple of weeks before Woody and Bob landed. He had to read about the completion of the flight in the local newspaper.

Jim has been vitally interested in aviation since he was a young boy after seeing Barnstormers in Alamogordo, NM. He was an aircraft mechanic in Naval Aviation aboard the USS Oriskany during the Korean War before attending the University of Arizona where he earned a degree in mechanical engineering. He then worked at the Yuma Proving Ground for almost thirty years, retiring in 1988. The first 20 years he managed to stay in the aviation field as Chief of Air Delivery. Some of the testing involved recovering crashed and disabled aircraft. They used an Aircraft Recovery Kit, enabling them to pick up airplanes with helicopters and fly them out. They recovered crashed military aircraft in several states and in the Philippines. He personally has recovered crashed aircraft around Yuma and in Mexico. The last seven years, he was Chief of Munitions and Weapons, responsible for testing of munitions and weapons ranging from the 60mm Mortar up to the Navy 16" Gun.

Jim is a member of the Experimental Aircraft Association, the Arizona Pilots Association, and is a mission-rated pilot with the Civil Air Patrol, among other things. He has restored several old airplanes. He used to have an Aeronca Sedan just like *The City of Yuma* and dreamed about finding that airplane, bringing it home and restoring it.

He flew sailplanes with Woody Jongeward "back in the Seventies," and now one of his favorite hobbies is ballooning.

He holds a commercial pilot's license for balloons and sailplanes and a private pilot's license with single-engine land and instrument ratings.

Jim and his wife Karen, have lived in Yuma most of their adult lives. Jim has three children, James Jr., Sandra Patane and Scott, four grandchildren and two great-grandchildren.

Preface - Shirley Woodhouse Murdock

The story of the Yuma endurance flight of 1949 has remained an important memory to me. It was exciting having my brother, Bob, involved in a flight that attained world-wide publicity. I was away at college in Colorado after the fliers had been in the air only about two weeks on their third and successful attempt to break the world's record for sustained flight. As they approached and then broke the previous record of 1008 hours, I began to see them in newsreels whenever I went to a movie in Denver.

Originally, in the mid-1980's, I was beginning to write a book of regional history—Roll, Arizona and the Mohawk Valley—and stories from our parents' pioneering days and their flying experiences, thereby expanding into a chapter about the endurance flight, because it was a Woodhouse family flying story.

Then, about 1997, as the fiftieth anniversary of the flight approached, and *The City of Yuma* was found, purchased by the Yuma Jaycees, and brought back "home" for restoration, I began to gather stories of that enterprise. It soon became obvious to me that I should expand that original "chapter" and write about the people who were spending hundreds of hours, and in a few cases, thousands of hours, restoring the airplane and showing it around town at major events, to raise funds for the restoration. It was necessary to pick the brains of the people who were working on those endeavors. Judy Griffen Spencer and her husband, Ron, were heading up the group to revive Yuma's memory of the flight. Ron kept a log of the group's activities. Then Jim Gillaspie, who is the kingpin, or "Technical Advisor," in the restoration of the airplane, was able to educate me regarding that major work. I had a dozen questions for him every week and he was able to explain the various processes to me, much as an instructor would to a student.

One day, after I had been writing and rewriting for weeks, Jim handed me a few pages that he had written about Yuma's faltering economy in 1949 and why something needed to be done to attract attention to the region and improve that situation——the reasons for the endurance flight. He offered those pages as an introduction, if I wanted them. I was pleasantly surprised by his writing and wanted to use it as the beginning of this book. I then turned to him to edit my writing, which he did with tact and diplomacy, but resulting in considerable "tightening up," condensing, and increasing clarity. I realized that his more objective overview of every aspect of the story would be effective in a book. He understood more about what was being done, and why. He's a historian and enjoys researching and writing, particularly about aviation history. Finally, I saw that this book would be at least half his work and presented the idea to him, that he is really a co-author. I said, "You're going to be recognized as co-author inside; you might as well have your name on the cover." Only after a repeat of that offer did he accept the title of co-author. With Jim's help, this book has become a more "polished" product.

Horace Griffen, "Griff," our dear friend of several decades, has answered at least a hundred questions from me, over the years, originally by phone or in person, then by frequent e-mail messages. He has edited everything that I have written. His memory and that of my husband, George, have helped me to bring back the original flight and the community spirit of Yuma in my mind.

Ron and Judy Spencer also answered many of my questions about the present-day activities——finding the airplane, bringing it back, then the promotions and fundraising for the restoration. Ron came to my home in Roll twice, with his log book, to confer with me for several hours. Judy came with him the first time. They also spent a number of hours editing some of my writing, sending e-mails and providing photos.

My appreciation extends to my family as well as the above-mentioned people. Big thanks to George for never complaining about the amount of time I spend in writing and rewriting or in research and interviewing people, either in person or by telephone. Also, the others in my family have been patient with the fact that I've had less time for activities with them.

It has been a pleasure to talk with a number of old-timers or their sons and daughters. I've had to ask some of them to relive the pain of losing their fathers, but they've all come back with amusing, poignant stories. There is still a great deal of appreciation for what those pilots, refueling crew members, and hundreds of other supporters did for the city of Yuma fifty years ago. It is hoped that, through this book and the activities of the present-day workers, that appreciation will enfold many more Yumans.

— Shirley Woodhouse Murdock

Preface - James A. Gillespie

Endurance is defined as (1) the power of enduring, specifically (a) the ability to last, continue or remain, and (b) the ability to stand pain, distress, and fatigue. When one looks up the word "endure," one finds that it means, "to hearten the heart." It is quite one thing to endure when an unfortunate situation occurs, but it is another to volunteer to put oneself in harm's way. Every one of these words directly relates to what Woody and Bob had to do for no greater individual gain than to be able say, "We hardened the heart, we persevered." This is the stuff of heroes. Although they didn't realize it at the time, and maybe not even today, the pain, distress, and fatigue that they placed on their bodies are still affecting them today. To touch on one area, the act of flying in a light aircraft without proper ear protection is detrimental even for just a few hours, but 1124 hours is off the chart. Woody, now will not attempt to talk on the phone because of his bad hearing and Bob has his problems too. People today think they are being abused if they have to stay in a small plane for five hours or a commercial airliner for 15 hours. When one thinks of physical abuse, Woody and Bob, between them, lifted over 52,000 pounds of fuel into that airplane while leaning halfway out of the airplane in the slipstream. The stress of flying at night, ever alert to the sound of the engine and thinking, "What do I do if the engine quits, where do I try to put it down?" The refueling runs, as beautiful as they were to see, were potential sources of danger. They were not just gently floating with the winds without a care!

I was attending Yuma Union High School when the flights were going on. At that time, I had no appreciation for what the pilots and all the support people were going through. On the surface, it sounded like one big party, something you could do every day. It was only after I started to fly that I learned to appreciate what they did. Almost every day, I learn something new about what happened back then. I marvel at how so many Yuma people could direct so much energy to one project for such a long period of time.

The story had to be told. It has been 50 years in the making. When I heard that Shirley was writing a book, I thought that was great news, at last all this effort will be documented. When she asked me for an interview, I jumped at the chance. I would be involved. When she asked me to be a co-author, I hesitated, but not for long. I couldn't believe she meant it. Deep down inside, I wanted the privilege.

We have written this book, not only to honor those of the past, but for the generations to come so they can have one more reason be proud of their heritage.

— James A. Gillaspie

Contents

Chapter One

The Roots of Aviation in Yuma, Including Endurance Flights

A lthough not widely known, Yuma has a history of endurance flights dating back to 1911. Yuma was the take-off point for a flight by Robert G. "Bob" Fowler that set the "world's record for duration and distance covered" in 1911. Bob Fowler had entered the First Transcontinental Air Race, in which William Randolph Hearst's *American* offered $50,000 to "the first man who could fly across the American continent within thirty consecutive days and complete the journey by October 10, 1911." Fowler's airplane was a Cole Flyer, a Model B pusher-type biplane built at the Wright Brothers' plant in Dayton, Ohio, where he went and had some three hours of instruction and ordered the airplane shipped to San Francisco. He had many troubles along the way, including a crash in Colfax, California and one in the snow at Donner Lake in the Sierra Nevada Mountains from which the Flyer was shipped to Los Angeles by freight train; it was repaired and he started all over again. On that second attempt, he took off from Ascot Park in Los Angeles on October 10th, exactly 39 years before *The City of Yuma* landed after its long flight, and the very day that he had hoped to reach the east coast. More delays occurred, one caused by Los Angeles fog which forced him to land at the Tournament of Roses Park in Pasadena, at night, then a "blown-up engine" at Banning, California.

Finally, on October 24th, he took off from Mecca, near the Salton Sea and arrived in Yuma on October 25th. His arrival was expected. Soon a small speck was seen over Pilot Knob; it grew to the size of a bird and then a full-sized airplane while over 2,000 breathless spectators assembled in the Yuma ball park and watched the wonderful object circle the enclosure and make a graceful landing almost on home plate. Bob was the first birdman to visit Yuma; the Cole flyer was the first flying machine in the city, and the first airplane to enter Arizona under its own power and not riding a boxcar. "Everyone from all over the valley was there," wrote Madeline Spain, who witnessed the great event. "We had been waiting for days, and the birdman's arrival was real exciting." Fowler then clattered into Maricopa on October 29th and landed alongside the railroad tracks across the street from the main post office. It had been a nonstop flight from Yuma——four hours and twenty-six minutes——and the 165-mile trip won him a world's record for duration and distance covered. Ruth M. Reinhold related this in her book "Sky Pioneering."

In June of 1929, former residents of Yuma, Martin and Margaret (Peg) Jensen, along with William Ulbrich, attempted to break the world's refueling record of 172 hours, 32 minutes and 6 seconds. This record attempt, made in a Bellanca aircraft flying over Roosevelt Field, NY, ended after only 70 hours, due to a fuel problem. The existing record, established at Ft. Worth,Texas, was held by Reg Robbins and James Kelly.

Martin Jensen and his wife were quite well known in Yuma, having been the subject of various headlines in the local newspaper. They were married in a Jenny while flying above Yuma in 1925. Martin was an aviation barnstormer, having come in second in a race to be the first to fly to Hawaii. In 1927, he was involved in an accident, described by Ruth M. Reinhold in her book, "Sky Pioneering", as being the most bizarre incident in Arizona's aeronautical history——a plane crash with "Leo," MGM Pictures' African lion mascot. Martin, flying a specially modified Ryan B-1 Brougham with Leo aboard, crashed in Arizona's Mazatzal Mountains and both survived. The lion was being flown across the country, as a publicity stunt for MGM, with planned stops all along the way.

Father of Yuma's Jackie Griffen, Ham Eubank, and another cowboy, Lewis Bowman, were working cattle in a rough canyon in the area when Ham's horse shied and started bucking. Ham got off and looked around, to determine what frightened his horse, and discovered the wrecked airplane with Leo in it. The pilot had walked away, hiking for 2 1/2 days in rough country before reaching a ranch. The area known as Hell's Gate was some of the most remote and isolated in Arizona. A group of intrepid cowboys improvised a sled from a forked oak and, using frightened mules, eventually rescued Leo from the area. He was transported to Phoenix by truck and to Los Angeles by train. The story is detailed in Frank V. Gillette's book, *Pleasant Valley*.

In 1949, the Yuma Jaycee-sponsored world-record endurance flight touched the lives of most, if not all, the 9,000 people of Yuma. For those directly involved, it was the adventure of a lifetime, something to be proud of, and not to be forgotten. As the years rolled by, memories faded and the stories took on new meaning. Among those old-timers, Bob Hodge loved to tell amusing stories about the flight around the campfire to anyone who would take the time to listen. Horace Griffen's main topic of conservation other than golf, concerns the flight. He has made numerous public appearances in which he described details of the flight. He has also kept in contact with most of the principal players, and was instrumental in putting on a fortieth anniversary celebration of the landing in 1989. He didn't know it, but he was destined to play an important part in the continuing story of Yuma's endurance flight.

FLY FIELD

Yuma International Airport, as we now know it, can trace its roots to 1925. This was the year that key individuals with a keen vision for future air transportation were appointed to the Yuma chamber of commerce aviation committee. This committee, chaired by Judge Peter T. Robertson with Everett Johnson and E. C. Briegger, went to work to secure an airport or landing field for Yuma. In approximately two years these men were able to: secure certain appropriated land (40 acres) from the government and have it cleared and leveled; secure the aid and influence of the Valley Country Club; secure the offer of a $32,000 hangar delivered F.O.B., Yuma; and secure legislation authorizing the board of supervisors to make expenditures in the establishment and maintenance of public aviation fields. Through their efforts, the 40 acres of sand was officially designated as an active airport. This designation was made possible through the efforts of the committee assisted by Col. Ben Franklin Fly. The airport, Fly Field, was subsequently named after Col. Fly, to honor him for his efforts to bring about the Yuma Mesa Project. In fact, he was often referred to as "The daddy of the

Yuma Mesa Project." In an April 28, 1928 issue of *Politics*, the Colonel, who initiated the legislation in Washington and carried it to a triumphant conclusion, was dubbed the "premier of all parliamentary solicitors." His efforts allowed for an unfailing supply of water to 130,000 acres of land and the initial expense of 12 1/2 million dollars involved in the construction of the Laguna Dam.

Fly Field was a start, but it had many shortcomings, one of which was the loose sand, and there were no facilities. In fact, Lt. R.N. Goddard, a flier from Imperial, California, in an address to the Legion in 1926, declared that most of the fliers from San Diego and other aviation fields passed up Yuma when flying over, due to the soft sand which could cause the airplane, when landing, to flip upside down. This and the fact that, in September, 1926, a New York aviation company designated Yuma as one of the line ports for a New York-to-San Diego mail route, was not lost on the chamber aviation committee. The committee felt that they needed at least 160 acres and some of that land would have to be bought. The rest would come from a trade with the government and a second trade with the Yuma Valley Country Club. The government was willing to go into negotiations because of the importance of having a safe and adequate landing field at Yuma. In fact, the Second Division Air Corps, Fort Sam Houston, on behalf of the government, offered to give to Fly Field a steel frame hangar capable of housing 12 airplanes. The Country Club was willing to trade some of its property for adjoining land that had been set aside by the government. The chamber named a special committee to work with the aviation committee to put through to fruition a first-class landing field in Yuma. This special committee consisted of: E.F. Sanguinetti, Norman Adair, Bert Caudry, Dr. H.D. Ketcherside, L.W. White, C.E. Potter, A.N. Kelly, Joseph Corey, William Wisener, F.W. Cresswell, and Frank Elliot.

The committee looked to the Yuma County Board of Supervisors to make appropriations for acquiring the rest of the land. The supervisors were authorized, by House Bill No. 51, to make expenditures for establishing public aviation fields. This bill, which was passed and signed during a prior legislature, was introduced by Representative William Wisener, of Yuma County, in the interest of the development of Fly Field. The board subsequently set aside $7,000 for improvements to the airfield.

Judge Robertson had lobbied for over ten years and carried on correspondence with key government agencies to keep alive the theme that Yuma belonged on the aviation map. The fruits of his labor were starting to pay off. The aviation commitment had received official recognition from the War Department. In the *Yuma Morning Sun* newspaper of November 13, 1927, it was announced that the government allocated 640 acres of land, to be used for a local flying field, to the aviation committee of the Yuma Chamber of Commerce. The article further stated that the new field is about four miles south of the center of the city on the main highway to Phoenix. Fly Field, consisting of 40 acres, named for Col. B.F. Fly, was abandoned and the name transferred to the new field. Work started the following day on the clearing of 160 acres. During the last week of December, 1927, Lieut. Harry Waddington, who for some time worked with the aviation committee, made the first landing on the new field.

On December 15, 1927, Congressman Douglas of Arizona introduced a bill asking for the lease of 640 acres of government land to Yuma County for 20 years at a cost of $1 per year. Simultaneously, a bill known as the Ashurst bill was working its way through the senate. This bill, if passed, which was drawn up by Judge Robertson, would allow all counties in Arizona equal treatment by the government, i.e., be able to lease government land for airport purposes and at the same rate.

People in Yuma were elated. Aviation officials predicted that, as a result of the passage of this action, it wouldn't be long until a postal route would be established between San Diego and Dallas. This was based on the fact that the southern route offers ideal flying weather with little mountainous terrain.

The next course of action for the committee was to urge early acquisition of two emergency fields in eastern Yuma County, one at Tacna or Wellton and the other at Aztec. At that time it was necessary to have an emergency field every 50 miles of an airmail route.

President Calvin Coolidge signed the Yuma Aviation Bill on February 27, 1928. Terms of the lease were for 20 years at $1 per year with privilege of renewal for another 20 years at the same rate. A special act of congress was necessary to grant the Secretary of the Interior authority for leasing the land. The act provided that the United States agencies shall have access to the field and that, in times of emergency, the government shall have the right to assume control.

Almost immediately, the aviation committee started lining up activities for the airport. Yuma was picked to be a night stop on three men's classes of transcontinental air races from New York to Los Angeles, and an international air race from Mexico to Los Angeles. These airplanes were expected to arrive in Yuma starting September 9, 1928. The chamber agreed to provide $1000 as lap money and provide free gas and oil to the racers estimated to cost $2000. Yuma was also picked to be a stop-over for the first All American tour of 25 airplanes and approximately 206 people. In June of 1928, it was announced that the United States meteorological and aerological station would immediately be constructed at Fly Field at a cost of $30,000. This station was to be manned by four Army personnel.

In 1929, Yuma was selected as the first stop for the women's transcontinental air race. Amelia Earhart had problems on landing and nosed her aircraft over, which destroyed her propeller. A new propeller and mechanics had to be flown in from Los Angeles to repair the aircraft so she could continue in the race. The ladies were not happy with landing in Yuma because their prior stop was in San Bernardino and the next stop for the night was in Phoenix. The city had to pay a significant amount to get the racers to stop in Yuma.

Development of Fly Field, and aviation in general, was slowed due to the depression, so activity at Fly Field was relatively slow until 1940. It became obvious that the United States was going to be involved in a world war. Part of the ramp-up to war status involved government-sponsored civilian pilot training programs and the build-up of facilities capable of warfare. Production of warplanes started increasing, in fact it was reported, on August 1,

1941, that 11,647 warplanes were built in the past year. The War Department needed facilities for training combat pilots and crews. Planning for the Yuma area started as early as 1939 when an aerial tour was made of a potential bombing (aerial gunnery) range to be located between Yuma and Gila Bend south of Highway 80. The government didn't make the announcement of this range until September 10, 1941 at which time the Yuma County Supervisors recommended that the Army consider Fly Field as an Army Air Corps base. Money for expansion of Fly Field as an emergency defense measure started pouring in early in 1941. This was the start of the takeover of Fly Field under a pending emergency situation. The Yuma newspaper of April 23rd reported that funding totaling $781,000 was received for immediate expenditure. $151,000 was provided by the Civil Aeronautics Administration (CAA), $420,000 under the Work Project Authority (WPA), and $210,000 through the Federal Priority Board. In August, 1941, another $635,000 was made available for re-paving the N-S runway and taxiway.

It was also in April, 1941 that General George Marshall announced that 500,000 civilians would be enrolled as volunteer observers to warn against aerial invaders. Yuma volunteers, under the control of the 4th Interceptor Command, Riverside, California, lined up to man the aerial observation post located on the water tower at Prison Hill. Their job was to spot and report all aircraft flying in the Yuma area. The major concern was enemy aircraft flying up from Mexico. In May of 1941, a plan for an Office of Civil Defense was announced and the first order of business was a call for volunteers. The office officially opened for business on December 22, 1941. One group of people formed a local glider club as an important part of national defense. Dick Haile, Leonard Jones, and Nick Wavers started building a BG-7 glider for training purposes. Sheriff Pete Newman, in November, 1941, formed a Sheriff's Air Patrol for the following purposes: to assist in locating lost airplanes; in the event of war, maintain patrol of power plants and lines, dams, and bridges. This was billed as the most advanced step in law enforcement since the installation of the two-way radios in Yuma County. The following were appointed to the patrol: Shields H. Craft, Captain; Denny Wraske, Lieutenant; Ralph Dusenberry; Jerry Nunnaley; Alfred T. Morgan; Kaleel Mittry; J.E.

Thomkins; Mrs. D. Wraske; and 18-year old Ty Hemperly. Shortly after this group was formed, Carl Knier, Airport Manager of the Phoenix Sky Harbor Airport was appointed Wing Commander of the Home Defense Aviation Air Patrol. Captain Craft wrote a letter to Knier offering services of the fully-organized Sheriff's Air Patrol. Thus the Yuma group theoretically became the first squadron of the Civil Air Patrol in Arizona.

It wasn't long after war was declared, that civilian aircraft was banned from flying within 150 miles of the Pacific Coast. Fly Field fell just inside the eastern boundary. Therefore, in order for local pilots to keep flying, they moved their aircraft outside of the boundary; most of them moved to Wellton. In February, 1942, it was announced that a new airport was being built ten miles east of Yuma consisting of two runways 3000 and 2400 feet long. Shields B. ("Bud") Craft, who had recently been appointed Vice Commander of the Civil Air Patrol, secured permission and funding from the CAA and the West Coast Command. This airfield was subsequently taken over by the government, which called it Auxiliary One.

Work started in 1941 at Fly Field and was completed in January, 1942. Two paved runways, 4200 feet x 150 feet, were completed. Yuma County had to maintain the field which was to be used exclusively by Army and Navy aircraft since local aircraft were grounded. In June 1942, the War Department announced authorization to spend $3,000,000 for the Army to construct an Army Air Corps Training School at Yuma. In July, 1942, Captain Barry M. Goldwater announced to a group of Yuma businessmen that the Army Air Base at Yuma would be one of the largest in the nation and within a year there would be a force of troops and civilian technicians numbering several thousand. In July, 1942, Del E. Webb Company of Phoenix was awarded the contract for construction of the training school. In December, 1942, the Aerial Gunnery Range opened east of Yuma and south of Highway 80. The first class of cadets arrived in January, 1943.

After the war was over, Yuma Army Air Field (YAAF) was scaled back and it was declared surplus in September 1946. The Fly Field portion went back under the control of the County, which again called it Fly Field. The government maintained control of the military side (South), however local people utilized facilities such as hangars. A few temporary buildings were moved to the civilian side (North) and used as a terminal for a number of years. In 1946, a number of airports, i.e., Marsh Downtown, Sturdivant's Somerton, Spain and Mellon's Yuma Air Park were established as training schools to take advantage of training pilots under the GI Bill. These airports offered competition to Fly Field so there was not a lot of activity there.

PILOTS WOODY JONGEWARD AND BOB WOODHOUSE WITH THE CITY OF YUMA, AT MARSH AIRPORT, JUST BEFORE FLIGHT.

THE CITY OF YUMA FLYING AT 1000 FEET, OVER THE COLORADO RIVER, HEADING NORTH, LOOKING BACK ON THE CITY OF YUMA.

This changed in April, 1949 when *The City of Yuma* airplane endurance flights started using Fly Field as a refueling site. The aircraft was refueled from a Buick convertible driving down the runway with the aircraft flying in formation. This feat was successfully carried out more than 1500 times without mishap. The objective of the endurance flights was to call attention to the excellent flying weather in Yuma. Three attempts were made by pilots Bob Woodhouse and Woody Jongeward to break the record of 1008 continuous hours in the air. They managed to break the record and surpass it by 116 hours, thus setting the new record of 1124 hours. This promotion, which involved over 600 volunteers, not only succeeded in setting a new record but also received publicity all over the world. This was undoubtedly the most successful promotion with far-reaching results ever carried out in the state of Arizona.

In late 1950, the Air Force started finalizing plans to reuse the air base. In February, the County Supervisors entertained a plan to lease the county airfield as a civilian-operated military training base.

Later, the Air Force sent in two different teams, (the final one on May 4, 1951) from Washington and The Western Area Defense Force to conduct a survey to possibly re-open the base on a limited basis. It was only ten days later that Senator Ernest W. McFarland and Congressman Harold A. Patten announced that Yuma Army Air Field would be reactivated. Their announcement went on to say that the Chief of Engineers would acquire necessary facilities at Yuma County Airport for use as a staging base for gunnery training in connection with Air Defense Operations. This activity would not disrupt civilian flying. A right of entry was given to the U.S. Air Force in June, 1951 by the Yuma County Board of Supervisors to lands and certain buildings under their control.

The base was named Vincent Air Force Base in 1956. The Air Force transferred the Air Base to the Marine Corps on January 1, 1959.

Eventually, the civilian side was renamed Yuma International Airport.

Chapter Two

The Flight

I t started as an innocent challenge and through sheer determination and pure tenacity, it developed into one of the greatest and most successful promotions ever conducted in the state of Arizona.

The year was 1949 and the place was Yuma, Arizona. Like most cities in the state, Yuma was in the middle of the post-World War II decline. The town's industrial base dwindled drastically after the war with the closing of the large Army Air Base and the other Army facilities associated with desert training. The Army Corps of Engineers maintained their Yuma Test Branch (YTB) at Imperial Dam, on a reduced scale, but on January 22, 1949 their prime test area, the Gila desilting basin floor failed due to high water flow. This failure immediately caused workload effort to diminish. YTB was subsequently closed in January, 1950 due to a rift between the Army Corps of Engineers and the Bureau of Reclamation as to fault.

Construction was relatively weak in that only a few housing divisions had been built since the end of the war. That left agriculture as the prime industry with the Bureau coming in a distant second. Yuma was not a destination for tourism because not too many people outside the state of Arizona ever heard of Yuma. The Junior Chamber of Commerce (Jaycees) started holding a world-class rodeo in 1946 that was listed as one of the top ten in the nation, but that news was reaching only a limited number of people. Agriculture was a rising star, thanks to the Bureau's Lower Colorado Yuma Projects. These projects were gradually converting raw desert ground into fertile agricultural land. This highly-productive acreage located in the Yuma, Gila, and Dome Valleys, and in the Wellton area and the Mohawk Valley to the East, had the potential to grow just about any crop, including produce, as long as water continued to flow in the Colorado River.

This lack of a solid diversified industrial base was a major concern to Yuma businessmen and, as a result, they were keyed to finding ways to broaden this base in order to help Yuma prosper. They, individually and in concert with their business organizations, turned over many stones to try and promote Yuma assets such as sunshine, unrestricted visibility, and clear skies with limited success. Try as they might, nobody had come up with an idea on how to knock a promotional home run. This was about to change.

In late January, 1949, members of the Yuma and Parker Chambers of Commerce held a meeting in Parker, about 125 miles north of Yuma. A number of Yuma members attended this meeting. By luck, four local businessmen, Horace "Griff" Griffen, Woody Jongeward, Ray Smucker, and F. C. "Frosty" Braden traveled together in one car. After they arrived in Parker, Glen Strohm, a Parker businessman whom they all knew, joined them for a tour of the town since this was one of the scheduled activities. During the tour, the conversation turned to two adventurous pilots in California, who were attempting to break the world record for keeping a single-engine aircraft flying continuously for an extended period of time. Glen, Woody, and Griff had previously known one of the pilots, Bill Barris, whom they had met during the government's civilian pilot training (CPT) prior to World War II. Bill and his partner, Dick Riedel, were flying an Aeronca Sedan, AC 15 known as *The Sunkist Lady*, out of Fullerton, California, in their effort to surpass the world endurance flight record of 726 hours. The Yuma men noted that these two pilots were getting a lot of attention from radio and the newspapers.

Ray Smucker, who was known for his quick promotional mind and was manager of Yuma's one-and-only radio station, KYUM, spoke up, "You know, there is a lot of publicity in an endurance flight. We could have one in Yuma. We'd show the world that we have 365 flying days a year and get our air base reactivated." The seed was planted and ready to be watered.

Horace Griffen recalls that the gist of the conversation the rest of the day was just kidding about it. The word had spread quickly among the other Yuma delegates and they all had their opinions on the subject. Later, on the way back to Yuma, the subject came up again and Woody, normally a man of few words, stated in no uncertain terms, "O.K., Ray, you get the airplane and Griff and I will fly it!"

The challenge was made and the spotlight was focused on Ray.

Within three days of that first conversation in the car, Horace Griffen received a call from Ray Smucker. "Well, Griff, I've found an airplane and we're having a kick-off meeting at Pete Byrne's office today at 4 o'clock." Pete Byrne was an attorney in Yuma and the men realized that a legal mind would be needed.

Ray Smucker was a man full of creative ideas. He initiated numerous programs to benefit young people and others in his town, and the city named "Smucker Park" after him. He had a half-hour program on KYUM Radio every morning called "The Sunny Side of the Street." As people drove to work, they heard him expressing sympathy for his unfortunate old friends back in "North Overshoe, Iowa." He said that the only problem faced by Yumans was "Where can we go for our health?"

He was president of the Arizona Junior Chamber of Commerce and suggested that the Yuma Jaycees sponsor the endurance flight project. This was their head coach talking and to say that they ran with that ball and scored would be an understatement.

Horace Griffen owned a Buick dealership which required his attention to the extent that he declined the honor (or the task) of being one of the pilots. The still-small group of planners determined that Bob Woodhouse was a likely candidate to fly with Woody. Bob was parts manager at Griffen Buick and, like Woody, a former Navy pilot. He loved to fly almost more than he loved to eat. He accepted the responsibility after Horace agreed to continue his salary.

Woody owned an electrical repair business with his brother, Howard, who agreed to keep the business going. Woody was 31 years old and Bob was 26. Both were married. One can only imagine the reactions of Betty Jongeward and Berta Woodhouse to this "crazy idea."

As the meetings continued happening in Pete Byrne's office and growing in size, many people reacted initially with some shock and amazement. Their questions about the logistics of accomplishing such an incredible feat were quickly answered by those who had been thinking about it for a few days. The group of volunteers multiplied rapidly as businesses and individuals offered their support and their almost-endless hours of work.

The airplane, loaned generously by Claude Sharpensteen and Mickey Lorang of the A.A.Amusement Company, was an Aeronca Sedan AC-15, N1156H, a four-place airplane with a 145-horsepower Continental engine. Marsh Aviation made its hangar facilities at the downtown airport available along with its head mechanic, Paul Burch, who was considered the maestro of aeronautics in Yuma. He, Bill Wilcox, Dallas Hovatter and Eddie Mendivil quickly went to work modifying the airplane. The right front seat and the back seat were removed. Auxiliary fuel tanks were installed, as the original wing tanks held only thirty six gallons of usable fuel. Two tanks, with associated hoses and pumps, were installed in the fuselage, one vertical and one horizontal, in the baggage area. They took up all of the baggage compartment and a portion of the rear seat area.

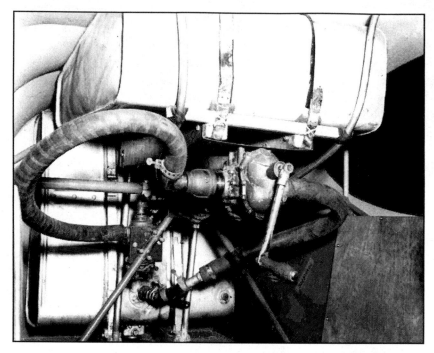

A VIEW SHOWING THE TWO FUSELAGE TANKS, ONE HORIZONTAL AND ONE VERTICAL, WITH THE ROTARY PUMP USED TO TRANSFER FUEL FROM THE 2 1/2-GALLON CREAM CANS INTO THESE TANKS. LATER, THE SAME PUMP IS USED TO TRANSFER THE GASOLINE FROM THESE TANKS UP TO THE WING TANKS.

The off-duty pilot used a hand-operated rotary pump to transfer fuel from cans into the fuselage tanks and later into the wing tanks. The addition of those tanks increased the total usable fuel capacity from thirty-six to over eighty gallons. Another system was designed and installed that allowed monitoring of oil quantity in the engine crankcase and a means to add and extract oil from the crankcase, thus, making oil changes available.

A VIEW, LOOKING FORWARD ON THE CO-PILOT'S SIDE, SHOWING THE BIG POT THAT ALLOWED FOR THE ADDITION AND EXTRACTION OF OIL TO AND FROM THE CRANKCASE. THIS ALLOWED THE PILOTS TO CHANGE THE OIL EVERY 100 HOURS.

Dallas Hovatter built a door that folded up and was fastened open, up and out of the way when necessary.

The plan was that the pilots would take turns at the controls in four-hour shifts. During the time that each man was not on duty as pilot, he could sleep, do physical exercises, and take care of the various chores that needed to be done to keep the mission on track.

The refueling car, a 1948 Buick Super convertible, provided by Griffen Buick, had belonged to George Murdock, Griff's service manager, who married author Shirley Woodhouse, Bob's younger sister, just two years after the endurance flight. George and Bob had initially practiced some refueling procedures with George in his Buick and Bob in George's Taylorcraft. That airplane held only two passengers; it was too small to be used for the flight and was too lightweight to fly in formation with a speeding car. As the plans for the flight progressed, Griff and George made a deal, involving George trading in that 1948 Buick convertible for a 1949 Buick Sedanette. Charlie Gilpin and Charlie Worthen, of Bert's Welding Works, (proprietor Bert Parrish) built a platform in the back seat area of the convertible with a protective railing around it. That platform provided a safe place for members of the refueling crew to stand when the car was speeding down the runway, as they handed food, fuel, and other supplies to the pilots.

Union Oil Company came into the picture as a major sponsor, providing all of the gasoline and oil for *The City of Yuma* throughout the project. The aviation gasoline was stored at Marsh's downtown airport where Bert Coffee filled the individual cans, filtering the gasoline through a chamois skin to remove any lint or other foreign matter. Then Chuck Mabery and Keith Smith ("Smitty") transported the full cans to the County Airport in a truck provided by Union Oil Company through Norman Bann, the local distributor. Some members of the refueling crew then transferred a few cans to the Buick convertible at the beginning of each refueling run.

The driver of the Buick accelerated rapidly to a speed of about 55 miles an hour——sometimes up to about 70——down a runway or across the triangular air base in whatever direction was into the wind. The entire paved air base was theirs to be used as needed. There was no military activity and very little commercial operation on the base; it belonged to Yuma County. Constant radio communication was maintained between the airplane and the Buick, by virtue of the expertise of Paul Estep, a radio specialist with the Civil Aeronautics Authority. Charlie Weeks was a radio man in the Navy and worked for A.A. Amusement Company, and he also worked on the radios during the flight.

The Aeronca, dubbed *The City of Yuma* and emblazoned with that name and the slogan, "The City With a Future, Yuma, 365 Flying Days Yearly," then joined up on the speeding convertible. The pilot maintained an altitude of only a few feet so that supplies could be handed up swiftly but carefully by the intrepid ground crew to the copilot of the moment. Both pilot and driver had to maintain an attitude of strictly business, dissuaded by nothing—eyes straight ahead, speeds synchronized exactly and in an unchanging direction.

They conducted a smooth relationship until the driver yelled "Geronimo" and the crew members sat down and braced themselves while the driver applied brakes with great determination to get that Buick stopped before it ran off the end of the 5000-foot runway into the soft sandy desert. The pilot had possible wind gusts to consider, knowing that his right wheel was near the spotlight of the Buick. It was also close to the head of his long-time friend, driving the car. The right hub cap of the airplane sustained several dents from bumping the spotlight during the flight, and the spotlight had to be replaced twice.

The off-duty pilot leaned out of the open door and down, in order to grab the handle of the can filled with life-blood for the Aeronca and bring it aboard.

EARLY MORNING REFUELING, ON FIRST ATTEMPT. WOODY FLYING, GRIFF DRIVING THE REFUELING CAR WITH BERTA AS PASSENGER. BOB HODGE HANDING BREAKFAST AND LUNCH TO BOB. PHOTO BY SAMMY WATKINS.

He was supported by a specially-devised safety belt which enabled him to use both hands to pick up the fuel cans. He then placed each can in the back seat area as quickly as possible, almost at the same moment that he reached for another, or for cans of oil, or their meals, water for bathing and shaving, clean towels, or changes of clothing. At times, even a high-speed kiss was exchanged between co-pilot and wife, as she stood tall in the Buick, supported or even boosted up by the versatile members of the refueling crew. Usually there were about twelve runs across the base for fuel and one for food, clothing, and other supplies.

During a normal single pass down the runway, four cans of gasoline, 2 1/2 gallons each, were usually handed up to the pilots. At the end of the runway the two vehicles broke formation, and the copilot transferred the newly-acquired gasoline into the fuselage tanks by using the hand-operated rotary pump as the pilot circled and prepared to join up on the car for another pass down the runway. They were then able to hand down empty cans as well as taking on full ones in subsequent runs. There were over 1500 such passes made by the time the endurance flight ended.

REFUELING CAR WITH SPUD PARKER, BOB HODGE, CHARLIE GILPIN,
PAUL BURCH, AND LOUIE MUELLER.
PHOTO BY ROD DALEY.

Two refueling crews were used; originally there were two refueling events each day. One was at 6:00 A.M. and the other occurred at 6:00 P.M. Horace Griffen drove the Buick for the morning refueling, with Bernie Pensky, Chuck Mabery, Keith Smith, and Howard Jongeward as the rest of the crew. In the evening, Charlie Gilpin drove the car, with Bob Hodge, Norm Bann, Louie Mueller, and Ralph Michaels handling the gasoline cans. Phil Neese was on "stand by." Betty Jongeward and Berta Woodhouse were always on hand, taking turns riding in the refueling car. Bill Linder, Floyd Estes, and Russell Phillips served regularly in various capacities.

"There was no committee chairman who assigned tasks, saying, 'You will do this, or can you do that? It all just fell into place,'" Griff mentioned.

Horace Griffen made his service department available, including George Murdock, his service manager. In addition to his duties at Griffen Buick, George was also tasked to keep the Buick in top running condition. Since cars of that day were not as reliable as today, this was almost a full-time job, especially toward the latter weeks of the flight. Due to the nature of how the car was being used, several sets of tires and front brake assemblies had to be installed to keep it safe. In those days, you couldn't get more than 10,000 miles out of a set of tires in normal driving conditions. Today, it is not unusual to get up to 70,000 miles. Fortunately for George, Goodyear Tire Company eventually took over the tire-changing responsibilities, and donated "a couple of sets of tires," as he remembers. Harold Sturges was manager of the Goodyear store. However, George had to handle all other aspects of keeping the car going. He had to take care of all driver complaints, in addition to keeping the car tuned to perfection. He had to tune the car twice a week, including changing the points and spark plugs. One day Griff said that the car just did not accelerate as well as it normally did. After running a series of tests to determine the cause, George decided to change the engine. In those days, Buick issued engines to dealers that were complete with all accessories. The engine would run in the box if you added a battery and fuel. This allowed George to be able to change out the engine overnight. He completed it at about 3:30 A.M., in time for the 4 A.M. refueling——another little drama within the bigger story.

Griffen Buick also provided a GMC pickup to run alongside the refueling car for photographers and other observers. Almost all of the newsreels and professional photographs were taken from that pickup during refuelings. Local professional photographers who took dozens of excellent photos were Emil Eger, Sammy Watkins, and Art Fszol.

Major sponsors of this venture were the Yuma Jaycees, A.A. Amusement Company, Griffen Buick, Jongeward Electric, Penn Signs, Marsh Aviation, Goodyear Tire Company, Union Oil Company, who provided approximately 9000 gallons of aviation gasoline and 200 quarts of oil, Continental Engines, who replaced the engine after the flight, and Tate and Hobart, the Shell Oil distributor, who provided the gasoline and oil for the refueling car—

no small matter. The entire city got involved: restaurants, a laundry, a sign company, a welding shop, service stations, petroleum distributors, and a long list of about seventy businesses donating support. Meals were prepared by a popular local restaurant, The Valley Café, owned by Whitey Stanton. They were picked up and delivered to the refueling crew by Harold Breech, chief of police, Jim Birmingham, Bert Power, police sergeants, and Ray Prather. Dr. Ralph T. Irwin was the official physician for the pilots. He wrote up their diets, low in fat, salt, and calories with some other restrictions. Both men lost several pounds during the flight. Woody weighed 133 pounds at the beginning of the flight and lost thirteen of them. Bob weighed in at 137 and lost eight pounds. They had no real problem with that; they spent a lot of time with their feet under a dinner table after the flight was over.

The slogan for the flight of *The City of Yuma* was "Ten-Ten!" Since the existing record, set by Barris and Riedel was 1008 hours, the Jaycees established 1010 hours as a minimum and "Ten-Ten!" was the call signal for the refueling car for their radio communication.

Twice the pilots had to give up and land because of mechanical problems, having failed to accomplish their goal. The first attempt began on April 21, 1949, at 4:51:50 P.M. and lasted only 74 hours.

Bob described what happened: "During the evening refueling run, the engine began to lose power, running on five cylinders. We were just goin' out from the airport after a run, and Woody was flyin' that time, and this thing is quiverin' and shakin' and backfirin'——one thing and another——and so we make a turn back, like you're not supposed to do, you know, but there are orange trees out there the way we were goin' and so it's better to get back to the airport if you can. If we'd had to land in those orange trees, we could have done that anywhere, but it would have damaged the airplane. Anyway, we got back to the airport and landed after a short conversation with Paul Burch. We decided to land and correct the problem before serious damage could occur to the engine. Then Paul and Bill Wilcox found the problem and it took a couple of weeks to repair and then we started again." They found that the original wing tank vents were not adequate for this method of refueling, which caused a vapor lock. This in turn caused the engine to overheat, burning a valve in one cylinder.

Bob and Woody flew to Fullerton, California in another airplane, to attend the landing of Barris and Riedel's airplane upon completion of *their* record-breaking flight. Bob's parents, Harold and Ethelind Woodhouse, flew to that event also in their own Cessna.

The second attempt of the Yuma pilots to break the world endurance flight record began on May 5, 1949 at 7:51 A.M. The pilots were forced to land again on May 12th after only 155 hours and 17 minutes in the air.

"That time," Bob said, "we started having a detonation problem again. I was flying it at the moment, and we had it on a high-power setting and, as we're goin' out, climbing out, very slowly—the thing starts quiverin' and shakin' and so I just started to turn——we were barely at the edge of the airport and I started to turn back around——of course, there was nothin' goin' on at the airport, you know, it's just about 500 acres of blacktop, and I'm tryin' to switch tanks, do anything, you can't do very much, you know for an airplane like that——if it's getting fuel, it runs, and if it doesn't, it doesn't. So, anyway, we made a trip around the airport, about ten feet high, in about a 15-degree bank, pulling carburetor heat and checking the mixture control and one thing and another. Nothin' did any good, but, about the time we started the second lap around the airport, we were flying in a trail of smoke. And, you know, it's pretty obvious where that came from, and about that time, there was a big explosion and it blew the crankshaft oil seal out of the front end of the engine and against the propeller. When you get a hole in the piston, then you build up too much crankcase pressure, then you've got a volatile mixture in there and so—— it wasn't too hard to figure out that it was all over right there. The windshield was all covered with oil. This was a problem that was insurmountable and we just rolled the thing level and landed. We all began to think about Smucker after that, how he got us into this deal. He was the perpetrator, for sure.

"The airplane had to go through a modification process and, in the meantime, Continental Motors had come out with a different crankshaft, a much better one, so the Jaycees bought that and I think they were in debt up to their ears, but we got the thing ready to fly."

Woody said, at the 40th anniversary party, "An interesting thing about the flight we did——one of the difficult things was cooling the engine. First of all, Paul Burch had rebuilt the engine and he put more clearance——he put a few thousandths more clearance between the pistons and the shoulder wall to make it run cooler and he was right——it worked. He also put some extra scoops out in front to make it scoop more air at low speed so as to cool it down. And it was a hot year. Our problem was to get to altitude without overheating the engine. We had thermocouples in all the cylinders and read-out gauges and monitors and after a refueling we'd slowly get up to about 8000 feet where it was cool and wait 'til it was time to refuel again. It was kind of a chore getting that done. I think that was one of the more interesting things about the flight. You think about the refueling runs being the most dramatic part of the flight, but actually, nursing that airplane up and down was a good share of it also. We did enjoy getting up to altitude——it's about 25 to 30 degrees cooler at 8000 feet and we were able to get up there where the air was clean and fresh, and besides, the guys in the refueling crews were passing a lot of gas."

It took months to get some of the parts needed. One of the modifications was the installation of a "sight gauge," showing the amount of oil in the crankcase.

The pilots, refueling crews, mechanics, and all of the city of Yuma were undaunted and eager to launch another effort.

The endurance fliers finally took off for their third attempt at the world record at 7:15:50 A.M. on August 24, 1949. Roy Slaten, manager of Western Union, was the official timer. Those wheels never touched the ground again until October 10th, 1124 hours, 14 minutes and 5 seconds—— almost forty-seven days later.

When the endurance fliers were just past the halfway mark toward breaking the record, on their 22nd day in the air, a third refueling was initiated. An article in *The Yuma Daily Sun* explains:

"At noon today, the ex-Navy pilots passed their 545th hour of continuous flying as they winged onward toward their goal of 1010 hours. Now more than half way through the proposed record-breaking flight, the pilots and crews are taking extra precautions to lessen the chances of anything marring a heretofore-perfect performance.

"In an effort to keep the plane's load as light as possible, the refueling operations will be strung out over a longer length of time. This morning, the refuelers started the operation a couple of hours early, dragging out at 4:00 o'clock instead of the usual 6:00. The darkness automatically slowed down the operation.

"Yuma Jaycees said today that they want to chill many rumors around town that the plane is having difficulty. The change in refueling schedule is more of a preventive move than a curative measure."

The time span through the night was thus divided up during the last three weeks of the flight. The runs for the 4:00 A.M. refueling, in total darkness, added another dramatic element.

4 A.M. REFUELING CREW. GRIFF AS DRIVER, BERNIE PENSKY, CHUCK MABERY, AND KEITH "SMITTY" SMITH. PHOTO BY EMIL EGER.

— 26 —

That heightened the interest and the concern around town. At that early hour, the pilots took on only enough fuel to last them until "the regular morning show," usually making three passes for fuel and one for breakfast and toilet articles. Horace Griffen drove the car for that refueling, as well as the regular morning run which was then changed from 6 A.M. to 8 A.M.

Even before the pre-dawn refueling was adopted, flare pots and lamps were kept burning along one runway every night. They had to be cleaned, filled, and lighted each evening, in case of a night-time emergency landing. Paul Burch's daughter, Shirley, says that some people have mentioned to her that they remember doing that task when they were in high school.

Bob explained to a group of interested Yumans recently, how the pilots first got "in tune" with the Buick for those passes. "Well," he said, "as the pilot, you're lookin' straight ahead, and you're seein' stuff out of your peripheral vision. We got where we could feel, when we'd start letting down to join up on the car, the right wing would get into some air that's comin' over the top of the car, so you'd have to roll a little right aileron into it to push that wing down. All of a sudden it's flying in air that's not coming straight to it, so you've got more lift. Well, you get a little deeper down in there, then some of that air that's coming off the side of the car begins hitting the tail, so you have to hold left rudder and right aileron—a cross-control situation. But then, as you go down into that disturbed air, you get into just the right position, which would be about when the right wheel is level with the spotlight on the Buick, then that sensation begins to go away, because that air is getting high enough above you. You could actually feel that, without lookin', you could feel that."

In describing how they learned the exact altitude to fly when joining up with the refueling car, Bob explained, "When Woody was flying, I'd put my hand up by the top of the instrument panel where he could see it. I told him: 'Just do what you're doin' regarding altitude, and if you're too high, I'll point down and you can work your way down; if you're too low I'll signal up.' So that's how we worked out that business of getting in the right position with the car."

He said, "It all went together very well. When we started out planning all this, we were gonna pick up those gas cans on a rope. Charlie Gilpin made a deal for us, out of a steel shaft about six feet long that had a hook on one end of it, and we had a cable on top of these gas cans and we were just gonna snap that into that hook and we'd pull the can up. Well, we got out there to practice with a can of water and Woody was flyin' and Griff was chasin' this shaft around that probably weighed about ten pounds or more, on the end of about 25 feet of rope, and you can imagine what that thing was doin' down there in that disturbed air from the airplane. So we picked up a can and it took almost a mile to do it. When I started pickin' that can up, it was pullin' the airplane down, so the can dragged along the runway and the bottom was torn out of it. So I just threw the whole thing away and I said, 'Woody, we're gonna get this stuff by hand. It may take a while.'"

Many people volunteered an incredible amount of time and energy and the refueling crew members' jobs were dangerous, but not one of them hesitated. It was considered an honor to serve on one of the refueling crews.

Griff says, "Our daughter, Judy, was born at 2:00 o'clock in the afternoon on September 12th and I was present for her birth, but I didn't miss a refueling run."

"About the only problems that we had on the third flight," Bob remembers, "were: the engine began to run rough after about three or four weeks. We called Paul Burch and he called the technician for Union Oil Company, Paul Goodwin, and he was in Alaska. He had an SNJ airplane that he used for transportation and he beat it right down here. I think Paul Burch came up with the idea. He sent us up a pint jar of water and a roll of masking tape and a couple of wrenches. He told us to take the vacuum line off the back of the manifold pressure gauge, and put a piece of masking tape over that line and then punch a pinhole in it, and then put it in this pint of water which causes some high moisture going into the engine. It was carbon buildup, is what it was. We did all of that, according to Paul's instructions. Griff was flyin' along under us in a Navion and he started seeing stuff come out——smoke and stuff, and then it started running clean, running better——we put the line back on the gauge and went ahead."

The pilots and refueling crews had almost no trouble with the weather. Charlie Gilpin, president of the Junior Chamber of Commerce, said that they had about three drops of rain one evening shortly before they landed, and that they experienced about 22 minutes of a wind storm one day. The Sun quoted the pilots as saying that every day was, "as we used to say in the Navy, the kind of a day when we wondered why anybody would want to stay on the ground."

THE CITY OF YUMA FLYING NORTH OVER FLY FIELD,
NOW CALLED MCAS YUMA, AND YUMA INTERNATIONAL AIRPORT

Bob said, "There was one time, however, when a dirt storm came up in the afternoon. We could see it comin' when it was down south of Mexicali. So we got on the radio and the car had a radio in it, and we called the guys and they rounded up the crew that filled the gas cans and they all came hustling out to the airport. By that time, the storm was right down this side of Somerton. It was a big wall of dirt. We used the diagonal runway because that's the way the wind was that afternoon and we made about two or three passes and the dirt was almost getting to us. The turbulent air that was ahead of the dirt had got there. We hit a pretty violent little whirly thing, and, if you'll notice, on the right-hand strut, it's wrapped with rope. There's a hunk of manila rope wrapped around the strut. Paul Burch had anticipated that, if we ever bumped the pipe on the refueling car and bent that strut, it might reduce its integrity and weaken the airplane, so he had wrapped that rope around it for a ways. It was a bumper, and we did bump it a pretty good lick that evening and it didn't hurt it.

"But then, we got out of there and headed out East, and slowly climbed away from it all. We'd taken enough gas to last us until the storm was over and then we came back and did the evening refueling before dark."

For the last three weeks of the flight, the Arizona network of radio stations broadcast a live interview with the pilots every morning. After the 8:00 A.M. refueling, the Buick was driven to KYUM where Ray Smucker, using the radio in the refueling car, talked things over with Bob and Woody and sent the conversations out over Arizona.

As they approached the existing world endurance flight record, they were in the national news regularly. KYUM was affiliated with NBC, and Morgan Beatty, on his *"News of the World"* program, mentioned "The Yuma, Arizona Endurance Flight" every evening at five o'clock, after the former record was broken. He would say something like "Let's see if those boys are still in the air, so now we'll switch to KYUM, 'The Hottest Little Station in the Nation.'"

Ray Smucker said, "We then brought in one of the pilots on the radio, so every night, either Bob or Woody talked on the radio, nationwide, and reported on how things were going." A clipping service sent articles from 32 foreign countries and they received letters from Germany, the Philippines, and Japan.

When they broke the record, the town went "bananas." The magic moment was at 7:15:50 P.M. on October 5th. The entire city went dark for one minute, prior to that time, by virtue of a master switch at the main power source for the street lights and residential lights. Then, at the appropriate second, the fire department whistle gave three short blasts. Factory whistles, train whistles, police, ambulance, and fire truck sirens, and thousands of automobile horns sounded. Yuma made itself heard and the lights came on again.

Near the end of the flight, the pilots and the airplane were becoming increasingly fatigued. The night-time refuelings were a source of some concern but they continued to be accomplished successfully. The pilots had planned to land on Columbus Day but their plans changed. On October 9th they flew over to Los Alamitos Naval Air Station in Southern California, where they took part in a "fly-over" as the opening event of an air show. The Navy had sent some officers over to Yuma to invite them to do that. Both Bob and Woody were in the Navy Reserves and the Navy's Blue Angels were participating in the show. Their friends, Barris and Riedel, former endurance flight record holders, were there as part of that show also. Horace Griffen, Betty Jongeward, and Berta Woodhouse flew to Los Alamitos for that occasion, where they were treated like visiting dignitaries.

"Well," Bob now admits, "The night before we went over there we were gonna have a morning refueling, and then fly over to Los Alamitos. Well, about midnight that night, we checked something that we did every hour; we checked all the cylinders. We had magneto switches, you know. So, on one mag, the thing was runnin' kinda rough. It's supposed to run on both mags, and it was fine on both, and it was fine on the left one, but on the right one, it was kinda quivering and missing.

"So, when I woke Woody up to take over at the end of my four-hour shift, I told him: 'I want you to listen to something.' So we did that check and it was shakin' on that one mag, but on both it was still pretty good.

"Woody said, 'What are we gonna do about that?'

"I said, 'Well, we aren't gonna check the mags any more 'til we get back from Los Alamitos!'

"Then, coming back out here over the sand dunes, we switched to that mag and it just quit completely. Of course, the prop continued to windmill, which sort of loaded the exhaust system up with unburned fuel and then, when we turned it back on, we got a big explosion. A big flash of fire came out of the exhaust stack. The bottom of the wing was polished aluminum so it lit things up pretty well. Anyway, when we got back to the air base, we told the refueling crew that we had something that we wanted them to listen to. Then we did that trick for them.

"Paul Burch came up with an idea real quick, as a remedy for that. He was gonna send us up a magneto and some wrenches and a hack saw and 'let's see, we gotta have some tin snips.' The plan was——and it wouldn't have been impossible to do this——if we wanted to continue, we would take the tin snips, cut a hole in the firewall, and then we would take all the wires off the magneto and set them aside. The magneto was bolted on with two nuts, so we'd take one nut off, and then we were gonna go to about 13,000 feet, shut the airplane off, get the prop to completely stop wind-milling, and then all we had to do ——and we would have had about thirty minutes to do this——well, the hacksaw was because there was a structural member that went across right behind where that magneto was, and you could get it off from the outside, but we were gonna have to cut that piece of the airplane out of there.

"All of that could have been done but, when Paul was tellin' us this, I was lookin' at Woody and Woody was lookin' at me, and we were thinkin' 'You know, we own this record; what are we tryin' to do?'

"So we voted Paul down. He didn't complain and neither did anybody else. We said, 'Well, what we're gonna do is this: we'll fly the rest of tonight and we'll refuel about every hour, a few cans of gas, and keep the airplane light so we don't have to use very much power, and then, any time, whenever——Smucker was the guy who was the contact man with the media——whatever time he wants this news release thing, that's when we'll land. That was critical, I guess, regarding the time of day, because of the national media."

The time for the historic landing was set at about 3:15 P.M. and the local schools all closed early so that the busses could take the children to the airport to witness the landing. There were fire trucks, ambulances, a doctor, and police officers standing by all night because the "Old Faithful" engine was running less smoothly during the refuelings.

Virtually all businesses in Yuma closed at noon and the crowd that gathered at the airport was estimated by officials to be between 12,000 and 15,000 people. Those officials said that there were about 6000 cars parked in the vicinity of the airport. When one considers that the total population of the town was only about 9000 people, the spirit that permeated the entire county becomes evident.

A PORTION OF THE CROWD, ESTIMATED AT 12,000 TO 15,000 AT THE LANDING OF THE CITY OF YUMA AFTER SETTING THE WORLD'S RECORD. NOTE THE SUNKIST LADY, AERONCA SEDAN 15AC AT FAR LEFT, WHICH HELD THE OLD RECORD OF 1008 HOURS. PHOTO BY ROD DALEY

When they were just about to land, Woody said to Horace Griffen on the radio, "Do you know what day this is? It's 'Ten-Ten'——October tenth!" Coincidentally, their landing date played right into that theme which originally meant 1010 hours.

Eight Navy F6F Hellcats came over from Los Alamitos, and formed an "aerial blanket" as *The City of Yuma* landed.

Bob said, "That was kind of a reciprocal deal for us having flown over there. But we had a sick airplane, and those guys in the fighter planes had their flaps down and everything and they're goin' around the airport and there was no way we could keep up with them. So we got on the inside of the circle and they were on the outside, makin' a bigger circle."

Then those military airplanes passed over the airport twice in a spectacular criss-cross pattern and a fireworks display was set off from a nearby hill by the police department. The roar of the crowd is still remembered by the old timers of Yuma. They were wishing that the pilots could hear them.

Bob said, "We had the guys in the refueling car come along and punch the tires and make sure we didn't have a flat, because we might have had a little trouble on landing if we'd had a flat tire. The tires were pretty soft and that shows up in the movies. The tires are makin' a kind of wide footprint there."

Bob says, "The time of the landing came on Woody's shift. He was a little worried because we hadn't landed for seven weeks, and we had knocked a spotlight or two off of the side of the Buick and bent the hubcap all up on the airplane. It wasn't really a problem, but we decided that I was gonna look at the ground real hard and tell him if he's three feet high or two feet high or whatever. So when he comes in to land and I'm lookin' at the asphalt which is goin' by about sixty miles an hour. It's kind of hard to tell just how high you are.

"Woody says, 'How's that?'

"I said, 'It's too high,' so he comes down about a foot, and we made a couple of adjustments like that and it got down to where it looked pretty good to me and I told him to go ahead, cut it off and land it. So he closed the throttle. Well, when he did, he began to feather the nose up a little and the airplane started fallin'. We had been about three feet high. He said that he heard me say, 'Oh oh!' Well, I wouldn't have done any better."

Before stepping out of the airplane for the first time in almost seven weeks, the endurance fliers taxied around the field so that everyone could get a close-up view and pictures of *The City of Yuma*. The pilots had wondered whether or not they would be able to walk. They were, although somewhat hesitantly at first.

OCTOBER 10, 1949. BERTA AND BOB WOODHOUSE, WOODY AND BETTY JONGEWARD WITH CONGRATULATORY HORSESHOE OF FLOWERS.

There was another celebration on October 12th. Pilots, refueling crews and other volunteers had been able to return to a semblance of their normal lives. The Yuma Union High School band led a parade, and the pilots and their wives sat up on the back of a new Buick convertible. They were presented with a gigantic horseshoe of flowers and were officially honored by their town and their state.

The dateline, "Yuma, Arizona" had shown up on metropolitan front pages and in major magazines and was heard on radios around the world, attracting attention to the 365 days of flying weather and the huge, inactive air base. On the day that the record was broken, *The Chicago Tribune* carried a front-page color photograph and attending article. *The Arizona Republic's* main headline on October 11th was "YUMA ENDURANCE FLIERS LAND," and the sub-titles were, "Record Set at 1,124 Air Hours," and "Huge Throng Greets Airmen; Both Appear in Good Condition." At the landing on October 10th, newsreel photographers were present along with many reporters, and network radio newsmen from near and far.

"The whole world heard about Yuma, Arizona," Ray Smucker said.

The pilots received congratulations from all over the world, but soon life for the major players started returning to normal with the exception of Yuma. As Ray Smucker had predicted, the attention given to Yuma during the flight started paying dividends as Yuma was named as one of ten potential sites for location of the U.S. Air Force Academy. On April 1, 1951, the Yuma Test Branch was reopened as a desert environmental test facility and renamed Yuma Test Station. The Air Force reactivated the airbase on July 7, 1951 as a training base. Yuma began a population explosion. It was and is, as proclaimed in the slogan on the side of that little airplane, "The City With a Future."

It should be pointed out that all of the work toward this project was in the nature of public service. Volunteers did it all. As Horace Griffen and Ray Smucker often say, "Nobody made a dime from the flight."

Chapter Three

The City of Yuma Rises From the Ashes

THE FINDING

The airplane, after being sold in 1953, was subsequently damaged in a ground loop while landing in Kansas. For some unknown reason, in July, 1953, someone requested that the CAA cancel the registration due to a wash out. In September, 1953, a request was made to the CAA to reinstate the registration, which they did. Many people in Yuma heard about the cancellation of the registration but not the re-registration. The rumor was going around for a number of years that 56H crashed and burned, totally destroying it. Shirley Burch and her sisters, Susie and Sally, grew up helping their father, Paul Burch, work on airplanes. They always pressured their dad to find 56H and bring it back to Yuma and fix it up. She remembers that Paul always said, "The airplane crashed and burned; it doesn't exist any more." In reality the damage was fixable, and 56H was returned to service and converted to a floatplane in 1978 when it was owned by Duane Cole. Cole also had a 180 HP Lycoming with a constant-speed prop installed. It went through a series of owners (a total of 19) until 1991 when Charlie Neal of Staples, Minnesota purchased it. It was about this time that Woody and Bob were inducted into the Arizona Aviation Hall of Fame and that it became generally known that 56H was alive and well. Gary Oden of McElhaney Cattle Co. in Wellton led the action to get Woody and Bob placed in that Hall of Fame. It was during the time leading up to their induction that Gary asked Bill Cutter of Cutter Aviation in Phoenix, to track the airplane down, which he did. The owner was contacted, but the asking price seemed a little high, so no further action was taken.

Also in 1991, Jim Gillaspie asked a friend, Dick LeMay, to look in the FAA registry to determine who owned 56H. He reported that Charlie Neal owned it. Jim's interest was to perhaps buy the airplane, return it to Yuma, restore it to the 1949 configuration, and re-introduce it to the people of Yuma. This plan was never implemented; however, in late October, 1996 another friend, Elmer Kettunen, called one day and asked Jim what he thought about bringing 56H back to Yuma. Jim told him it was an excellent idea and then asked if he wanted to know where it was located. That simple question started another round of togetherness between the people of Yuma and *The City of Yuma*.

Elmer called Charlie Neal to find out if he would sell it. Charlie replied that he had recently lost his Medical Certificate due to a brain aneurysm and yes, he would sell the airplane. He said that his son, Chris, flies it but he owns a crop-dusting service and he doesn't need to fly that airplane. Elmer subsequently called two other people, Horace Griffen and Orval McVey, to join the group. At the first meeting, everyone agreed that they should bring 56H back to Yuma and that Jim would head up the effort since he was the youngest. At a subsequent meeting, Horace Griffen reported that he had talked to his son-in-law, Ron Spencer, a member of the Yuma Jaycee Foundation, and asked him if the Foundation might be interested in this effort since the Jaycees were one of the original sponsors of the endurance flight. Ron replied that he thought they might be interested, but he needed more details before presenting it to the Jaycees. He felt that they would want to know how much money would be involved, i.e., what the airplane would cost plus an estimate of the cost to refurbish it. Jim Gillaspie was asked to come up with these numbers. Planning numbers were given to Ron and he started laying the ground work with other members of the Foundation.

THE RETURN

The original idea was to bring 56H back to Yuma and restore it as closely as possible to look like it did in 1949. After this was completed, it would be placed in a Yuma museum. This resulted in a long dialog that lasted several months between Jim, Mr. Neal, and his son, Chris. During the course of the many conversations and negotiations, it was made clear as to the reason for wanting this particular airplane. Chris stated that he had read the log book entries on the three endurance flights, but didn't realize that the record had been broken on the third flight. Chris was told that the

Jaycees wanted to refurbish the aircraft and return it to the 1949 configuration as closely as possible. Therefore, they wanted to buy the airplane without the 180 HP engine, mount, cowling, propeller, floats, skis and hardware. Then they would have to go and find the original type engine, prop, mount, and cowling to put on this airplane. Chris replied that he had another 1948 Aeronca Sedan, and that it had the original engine, mount, prop, and cowling and that they could have them instead. Chris described the engine as "high time," but he thought it only had one low cylinder. Jim then requested that Chris determine a price for what the Jaycees wanted.

Finally, on April 19, 1997, Chris called with a price of $19,000. The committee met with Ron Spencer and Garth Worthen of the Jaycee Foundation at Dakotas for lunch and the price and estimated costs were discussed. Ron and Garth thought the project was something that the Jaycee Foundation would be interested in, and they would take it to the Foundation Board. At this meeting, Ron indicated that now the Jaycees would be interested in returning 56H to an airworthy condition. Jim pointed out that this was a new ball game, that of returning it to airworthiness status, but one that could be solved with time and money. The question was raised that, since the Foundation is a non-profit organization, would the Neals be interested in donating the airplane? Jim was tasked to ask them this question. Ron and Garth asked Jim for his long-term commitment to help bring it back, and be involved in the restoration.

The price was given to the Foundation and they agreed to fund the purchase price as well as the cost to go get it. Chris was asked if they might be interested in donating the airplane to the non-profit Foundation and he said that they might be, but he would have to look into it. After checking with their accountant, the Neals decided not to take advantage of that suggestion. The price was further negotiated and reduced to $18,000. Everybody agreed to this price; however, there was five feet of snow on the Neal runway so there was no hurry to go get it. The Foundation established a committee of Ron Spencer as Chairman, Garth Worthen as Vice-Chairman, and Mary Worthen as Treasurer. Ron further appointed his wife, Judy, as Administrator. A $1000 retainer was sent to the Neals in May, 1997 with the understanding that the balance would be due after inspection of the aircraft. Ron, Judy, and other

Foundation members met with Wayne Benesch, Attorney, and Calvin Brock, C.P.A., on June 26th to seek their guidance on the legality of their plans for the airplane. The plan called for bringing it back to Yuma, restoring it, and eventually giving it to another non-profit organization with a home so that the airplane could be put on long-term display.

During the course of the negotiations, Chris offered to let Jim fly 56H to Yuma and then they would ship all the components back and forth until everything was in place. The airplane, with the 180 HP engine and 36 gallons of usable fuel, only had a safe range of slightly over two hours. At a speed of approximately 105 miles per hour, this meant that they would have to land every 200 miles or so and refuel, which was not desirable. Also, some Foundation members felt that there was a risk with flying it back. Ron and Jim investigated options on the most economical and timely method to bring 56H back. The first thought was to get a trailer and go get it. This was ruled out because of the time required and the potential of damage when transporting it on an open trailer. They also considered hiring a professional aircraft mover to pick it up and transport it to Yuma. This option was dropped due to costs and potential for damage. They looked into the option of renting a one-way moving van. After talking to Gary Magrino and Artie Durazo of Multi-Tech, and comparing dimensions of the disassembled airplane and the volumetric capability of a 24-foot Ryder truck, they considered that the best option so far. Artie gave them a few tips on how to economize the effort and, based on his recommendations, this method was selected. It seems that the cost could have been twice as high as that paid, but Ryder had an oversupply of trucks in the East and they wanted to get them west. Multi-Tech had a need for this size of truck, so everyone was happy. Ron and Jim developed a plan that called for hanging the wings on the side of the van and placing the fuselage in between, and the tail feathers, cowling, fairings, and other items on the floor underneath the fuselage. The biggest problem would be how to secure everything. There have been stories of how aircraft, carried this way and not properly secured, were pounded to pieces due to the relative light weight of the airplane and the stiff suspension of the truck. This plan also called for Jim and Ron to travel to Phoenix by shuttle van, then fly commercial to Minneapolis with tools and rent the truck there, and, after picking up the airplane, drive back to Yuma. One problem remained: where to put the airplane after getting back to Yuma. Jim, who had been keeping Bill Jewett aware of the progress of this effort, mentioned it to Bill, a big supporter, and he graciously volunteered his warehouse.

By this time, interest was building in Yuma. Reactions ranged from disbelief to pure excitement that indeed there was such an airplane. Ron had contacted KYMA and *The Yuma Daily Sun* and told them what was about to happen. They responded by doing interviews. Pam Smith of *The Sun* interviewed Ron and Jim for a story to be released just before their return to Yuma. She also contacted the newspaper in Staples and told them what was happening and requested them to photograph the disassembly and loading of the airplane.

Ron and Jim left Yuma with tools in hand on the eighth of July and traveled to Phoenix on the shuttle bus. From there they flew to Minneapolis. The taxi ride to the Ryder truck agency was pretty exciting. The driver got lost a couple of times, one of which made Ron and Jim believe it wasn't possible to "get there from here." Finally, the lost was found and they arrived at the Ryder Truck Rental Station. After renting the truck and padding, they proceeded to drive towards Staples, 120 miles away, via U.S. Highway 10. A few miles outside of Minneapolis, they hit a rough section of highway that caused them to question why they were there. The highway was so rough that the truck, heavily sprung and capable of hauling 26,000 pounds, could not be driven faster than 35 miles an hour. Ron said that they needed seat belts just to stay in the truck. Maybe the guys that had problems hauling aircraft that way knew what they were talking about. They persevered and continued on and stopped for the night in St. Cloud; enough for one day.

Up early the next day, they continued on. The plan was to arrive early that morning in Staples. Once in Staples, they were to call the Neals for instructions on how to get to their house. They lived eight miles west and north of Staples. Jim tried to call several times but got no answer. Well, "no problem," they could find somebody who knew them and get instructions to their farm.

Several businessmen were asked but nobody knew them. That's unusual. Later they saw a hardware store and immediately thought that they would know the Neals because all farmers need a hardware store. That may be true, but it turned out that this farmer evidently didn't need that store because they never heard of the Neals. This left Ron and Jim scratching their heads, but it wasn't a complete loss, because they were able to buy eye-screws, nails, and other items necessary to secure the airplane, if they ever found it. At this point, they were beginning to question if there was really a farm with 56H sitting on it. Jim remembered that, when negotiating with Chris, he had mentioned Verndale, Minnesota as being close to their farm, so the decision was made to drive to Verndale which was eleven miles west and north of Staples. They stopped at a convenience store on the east end of town and tried to call the Neals once again. Nobody answered the phone. They waited five minutes and called again; still no answer. Jim, callused and hardened by this time and expecting a negative answer, asked the clerk if she knew where the Neal farm was located. Surprisingly, she said, "Yes," and proceeded to tell them how to get there. But she got confused and again they wondered if they could "get there from here." Finally, Jim tried the phone once again and this time got Mrs. Neal who gave them clear instructions on how to get to the farm. Within ten minutes, they were driving down the long built-up dirt road leading to the farm house. By this time, and with all the delay and wondering, Ron and Jim were most anxious to see 56H. As they drove up to the farm, the airplane was not in sight, adding to the anticipation, but at least there was a small closed hangar there, so it had to be there. They knocked on the door and Mr. and Mrs. Neal greeted them, but Chris wasn't home and wouldn't be back for a while. Chris was to have taken the engine off prior to their arrival, so they expected to catch him working on it. The Neals inquired how the trip went. Ron and Jim described trying to reach them and Mrs. Neal said that she must have been outside when they called. Jim told about how they couldn't find anyone that knew them in Staples. Mr. Neal described how he had lived there all his life, having been born only a mile away. Ron and Jim wanted to see the airplane in the worst way, but they didn't want to seem too anxious. It seemed a long time before Chris returned and opened the hangar. There it was——a beautiful sight, even with the engine cowling off, and it was about to begin its journey back to Yuma. Seeing is truly believing. Jim quickly looked over the exterior, then opened the door and climbed in and sat there quietly savoring the moment for an extended period of time, unaware of what the other people were doing or saying.

Finally, Jim crawled out of the airplane, and then realized what had taken place. He had bonded with 56H and this was the start of a long relationship. Jim was soon to find out that he wasn't the only one or the first to experience this bond with this airplane. The first task was to verify that this airplane was, indeed, the airplane that set the record. The first clue was that there were holes in the windshield fairing that offered evidence that the two antennas, especially mounted for the endurance flights and shown in the 1949 photos, had been there. Other evidence found were patches over holes in the firewall in which oil lines were routed between the engine and the cockpit. These lines made it possible to add and extract oil from the crankcase. The log books were compared to repairs to fuselage damage and were found to be a reasonable match. The logs indicated that the airplane only had a total of 3100 hours flying time since it was manufactured in 1948. Bob Woodhouse and Woody Jongeward flew almost half of those hours. On their first, second, and third attempts, they flew it approximately 74, 155, and 1124 hours for a total of 1353 hours, plus the time that they spent experimenting with refueling techniques, deciding how to do that, then perfecting the technique. The airplane only had 137 hours on it when they started. After making sure that this was the original 56H, Ron and Jim moved it outside and put the cowling back on and took pictures for a before-and-after sequence.

56H AS IT LOOKED WHEN IT WAS PICKED UP IN STAPLES, MINNESOTA
JULY 8, 1997

They were in the middle of this when a reporter for the Staples newspaper showed up. Jim wondered how he knew where the Neal farm was. He said that the newspaper had been asked by Pam Smith to cover this event and he started taking pictures, and at the same time, interviewing anybody that would stop and talk to him. The Neals were in awe when the reporter showed up, as to the attention being put on them. They decided that this was such fun that Mrs. Neal called the Wadena, Minnesota newspaper and suggested that they should come out and cover these happenings. They did, so Jim and Ron had to divide their attention between disassembling the airplane and answering questions from two reporters at the same time. They started to take the wings and tail feathers off and Chris, his dad, and a helper started taking the engine and prop off. After the wings were taken off, they were loaded into the van and secured to the side walls with the leading edges down, lightly touching padding that was resting on the floor. In the middle of the afternoon, Mrs. Neal announced that she had cooked a farm lunch for everybody, including the reporter that was still there.

She had prepared ham, quiche, and various kinds of vegetables, baked goods, and even a rhubarb pie. Most delicious, that alone made the trip worthwhile. Mrs. Neal explained, "This is like our dinner. We have a light supper later in the evening." This lunch had a good effect on the reporter because he wrote an amazingly-accurate account of what he heard. After eating, Ron and Jim had to develop a way to load the fuselage into the truck. This was solved by driving the truck out into an alfalfa field and backing up to the built-up dirt road leading to the Neals' house. This placed the bed of the truck at about the same level as the road. A couple of 2x10's were used to bridge between the bed and the road. After the engine was removed, they rolled the fuselage out and down the road. At the truck, they turned the fuselage 90 degrees and pushed it into the van while, at the same time, forcing the landing gear together to fit through the door opening.

LOADING THE FUSELAGE INTO THE RYDER TRUCK, STAPLES, MINNESOTA, JULY 9, 1997

The tail feathers and other items were loaded under the fuselage. They used eye-hooks screwed into the wood floor and tie-downs to secure everything. They worked until approximately 8 P.M. They drove to Staples where they stayed all night, returning to the farm the next morning to load the engine and the remaining items. The last thing was to complete the paper work and pay off the balance.

Ron and Jim left the Neal farm at noon on the tenth of July. They took U.S. Highway 10 northwest to Fargo, North Dakota and then got on I-94 and drove to Wibaux, Montana. Along the way, the weather turned bad with thunderstorms, hard rain, and low, dark clouds. "One of them looked like a big, rotating tire, with bags hanging down out of it. They were all threatening," Jim said, and he was glad that they weren't out there with the airplane on an open trailer. It became a race to get out from under those clouds. They learned later that there was a tornado in the area, but they didn't see it.

The next day they continued west to Billings, Montana, and then joined up with I-90. They stayed on that freeway to Bozeman, then turned south on highway 191, passing through Gallatin National Forest, West Yellowstone, and stayed overnight at Idaho Falls, Idaho. When passing through Bozeman, they encountered more washboard roads in which they couldn't go faster than 35 miles per hour. Otherwise, the maximum speed that they could go was 65. They had to stop every few miles to inspect everything and make adjustments if necessary. The next day they continued south on I-15 and spent the night in Las Vegas. The following day they continued south on US 95 and returned to Yuma on Sunday, the 13th.

An amusing thing happened when they passed through the Arizona Inspection Station at Ehrenberg, Arizona. The inspector asked, "What you got back there?" Ron answered, "An airplane!" The inspector said, "What?" Ron repeated, "An airplane." The inspector looked kind of puzzled and said, "Oh well, OK," and waved them through.

The next day, they unloaded 56H and had an open house so people could see it.

UNLOADING 56H IN YUMA, JULY 14, 1997. RON GILLASPIE AND BILL COX.

RON AND JIM WITH THE FUSELAGE AFTER IT WAS OFF-LOADED IN YUMA.

The only damage, as a result of the move, was to the engine baffling. This was easily repaired. Pam Smith's article, *"The City of Yuma Flying Home,"* was in *The Yuma Daily Sun* on Friday, July 11, 1997. Judy Spencer called a lot of people to come and see it. Young and old alike came to see the airplane. Many old timers, people who were in Yuma during the flight, and their sons and daughters——even their grandsons and granddaughters——came and shared their pictures and talked about what they were doing during that period of time. One of those granddaughters was Marilyn Gardner, whose husband, Greg, of KAWC, has worked on the moving crew, whenever the airplane was moved around town to various events. Greg has been active with the promotions, also, particularly with his radio programs. Marilyn's mother, Betty, is the daughter of Ray Prather, who was involved in transporting the meals to the fliers in 1949.

Participating sons and daughters of those involved in the flight included Judy Spencer, daughter of Horace and Jackie Griffen; Nancy Woodhouse, daughter of Bob; Shirley Burch, whose father, Paul, was the mechanic; Perry Pensky, whose father, Bernie, owned

SHIRLEY BURCH WITH SON CHRISTOPHER, INSPECTING AIRCRAFT AFTER IT GOT BACK TO YUMA.

Penn Signs, which made the signs on the airplane and on the refueling car; Garth Worthen, whose father, Charlie, helped build the rack on the refueling car; and Claude Sharpensteen III, whose father was co-owner of the airplane and generously loaned it to the Jaycees. After hearing many stories that were told with affection about all the events that occurred during that period, Jim realized that it really wasn't necessary to reintroduce the airplane to Yuma. The simple truth was that 56H had never left Yuma because it had lived on in the hearts of all these people. One teenaged boy asked, "Why was the airplane allowed to leave Yuma?" No one could answer his question other than to say, "That's the way they did things in those days." Nobody believes it will leave Yuma again. Pam Smith came and, on seeing so many sons and daughters of the principal players in 1949 there, she dubbed the group, "The Endurance Brats."

THE RESTORATION

Work to restore 56H to its glory days was started a short time later. Many volunteers, at times perhaps too many, have come forward to help. All ski and float hardware was removed, instrument panel replaced, and fuselage and empennage prepped for paint. Shirley Burch was especially pleased to be able to use her father-taught skills on this airplane. Among other things, she enjoyed being able to remove a very large rotating beacon that was mounted on the rudder. That beacon was not on 56H in 1949. Shirley also worked on the fuselage, making patches where necessary after the removal of some of the hardware. Sun Western Flyers and Keith Tyree donated materials to repair the fabric. The aircraft was painted in the same color scheme by Jimmy Allen in Tom Pulda's paint booth. Jimmy Allen, a volunteer, is considered the best painter in Yuma. Ron Contreras of Penn Signs painted the lettering. Penn Signs did the original painting in 1949 and is the oldest licensed continuously family-owned business in Yuma. The wings were stripped of paint by Bob Schmidgall and then hand polished by numerous volunteers including the CAP cadets from Yuma Squadron 509 (see list on page 95).

Jerry McGuire, who owns a part-time business called J&S Sewing, has volunteered to rebuild and re-upholster the seats and side panels, using FAA-approved materials.

Since the airplane was to be flown again, it was necessary to give it a complete inspection to determine the exact condition. Leak-down tests were given to each cylinder and three were found to have little capability to hold compression. The engine had a record of being hard on cylinders. Since the engine was close to the recommended overhaul time, a decision was made to tear it down, this being the only way to determine overall condition. The engine was moved to Ernie Munoz's house where he, John Youkey and Jim Gillaspie took it apart. The general condition was noted and critical components were checked by micrometers to determine amount of wear. The decision was made to send the crankshaft, camshaft, rods, tappet bodies, crankcase, and other items off to be reconditioned. Jim called Greg Merrell, owner of Aircraft Specialties Services of Tulsa, Oklahoma, to tell him that the items were being shipped to him. At the same time, Jim told him about the famous aircraft that they came from. Greg responded by saying that he was interested in this and that they would give the group a reduced price on parts and some free labor. He also said that his company doesn't work on crankcases, but that DivCo down the street did. Jim called DivCo, talked to Charlie Jarvis, and found out that Greg had already called him. They also gave a break on the cost of reconditioning the case. In the meantime, Jim Siemens and Jim Gillaspie checked the cylinders and found that four of six were worn beyond limits and needed to be overhauled. The other two had already been bored out because of wear. Jim called DivCo, a company that also specializes in cylinder overhaul and found out that it is cheaper to buy new ones than to overhaul the old ones. Because of this, they contacted Continental Motors about getting new ones. That effort is described in the Promotions section of this book. All items were sent off, reconditioned, and returned. New bearings, gasket kit, rod bolts and nuts, and other items were purchased. When all components were returned, the engine was moved to Jim Siemens' workshop and reassembled by Ernie Munoz, Jim Siemens, John Youkey, and at times, Jim Gillaspie. The magnetos were relatively new and didn't require attention. The carburetor venturi and float were changed in accordance with FAA Airworthiness Directives. The generator and starter were disassembled, inspected, and tested with the help of Jim's Harley Shop and Howard's Alternator and Generator Service. The fuselage was moved to Jim Siemens' house to facilitate hook-up of the engine. The difference in size of the two engines has required replacement of certain items. Ernie Munoz, FAA Certified Mechanic with Inspection Authorization (IA), is heading up this effort.

The engine, mount, and cowling came from a different airplane, a Canadian one. Since all Aeroncas were individually built by hand, there were slight differences that had to be overcome. The design of the three-point engine mount made it difficult to get all three bolt holes lined up, which is normal. Jim Gillaspie and Bob Martin developed a way to properly line up the mount to the fuselage. The mount was later powder coated, which was arranged by Nick Curtis and accomplished by his sons at Powder Tech, in Phoenix. The cowling did not fit properly, so adjustments had to be made by drilling new holes. Two new lower cowl pieces were fabricated by Jim Allen and mated to the rest of the cowling. The cowling now looks better than when it was new.

The airplane has been completely inspected and corrections made. A new windshield has been installed as well as the side window glass. Carl Franks, a winter visitor from the State of Washington, made a fold-up refueling door similar to the one used in 1949 that was made by Dallas Hovatter. This door will be used when 56H is placed in a museum. The Goodyear brake system has been resealed by Clay Garrison, another winter visitor from Missouri, and Jim Gillaspie. Garth Worthen, in conjunction with Foxworth Lumber, replaced the rear wooden decking. H. G. Frautschy, editor of the Experimental Aircraft Association's *Vintage Airplane Magazine* (circulation 25,000) found out about this effort and offered his services. He wrote a brief article about *The City of Yuma* for the January, 1999 issue, and asked if anyone had certain items that they would like to donate. This included flying wires, an instrument light, and control wheels. Original control wheels were donated by D. Carlson of Hay Springs, Nebraska. Other items were contributed by readers in Rhode Island and Missouri. These items are rare and difficult to find.

Other contributors and volunteers donating time, energy, and material to the project include: Ray Williams, Western Representative for the Arizona Pilots Association, Gary Wedding, owner of the Avionics Shop at Eloy, Arizona, who provided vintage instruments, etc; Therole Miller whose pads were used for transporting the wings and various certified parts for the engine; Bill Wilcox, vintage radio parts and instruments for auction; Jesse Mooneyham, trailer and engine removal; Mike Thompson, YPG Heritage Center Curator, vintage items for auction; and Harold Gelman, who has assisted in locating engine parts and a propeller.

Work on the airplane and fund raising is continuing and 56H is expected to be airworthy in August of 1999. The airplane will be reassembled at Bet-Ko-Air, Yuma International Airport. Once the test flights have been completed, it will be hangared at the McElhaney Cattle Company in Wellton, Arizona. This was made possible by Gary Oden and George Murdock. *The City of Yuma* will be flown out of there until a permanent home is found in a Yuma museum.

Chapter Four

The Promotions

Return of *The City of Yuma* and its restoration efforts have been supported, and in many ways made possible, by the tenacity of the promotional group consisting mainly of Ron and Judy Spencer. Ron and Judy have worked in many capacities on the committee and it has taken them in various directions. These include presentations to civic groups, letter writing and hundreds of phone calls, a monthly newsletter, and scheduling numerous appearances of the airplane around the city, trying to get contributions, major sponsors, and a permanent home for *The City of Yuma*. Jim Gillaspie, Horace Griffen, Garth and Mary Worthen, Nancy Woodhouse, Shirley Burch, Jerry Barnett, and Greg Gardner have been involved in many of these efforts mostly in support roles. The two pilots, Bob and Woody, have appeared at some activities along the way, including barbecues and radio and television programs.

The group was searching for a permanent home for the airplane to be on display, even before Ron and Jim went to Minnesota and brought it back to Yuma. They went to the Airport Authority board on April 30, 1997 because they knew that a new terminal was going to be built.

"We told them that *The City of Yuma* was coming back to Yuma and it could be made available to hang inside the terminal if possible," Ron says. "The board members were interested, but they were afraid that it wouldn't fit inside the building." They didn't have the exact dimensions yet—the length of the fuselage and the wingspan. Later, on July 22nd, after the dimensions were known, they met with a member of the board with the same request, but after reviewing the plans, it was decided that the structure wasn't big enough to hang an airplane 26 feet long with a 40-foot wingspan. The board offered the group a plot of land in the employee parking lot, to build a structure to put it in. The group said that they would consider that. Later, they received an official letter from the Airport Authority, saying that they were sorry but they wouldn't be able to display the airplane in the new terminal.

"We then pursued another avenue," Ron says. "We went to Joyce Wilson, City Administrator, and talked to her. She said that she would approach the Crossing Park and the Madison Avenue Development regarding a place for *The City of Yuma.*"

A few weeks later, in September, Ron, Judy, and Jim met with Rafael Payan, Acquisition and Development Chief, of the State of Arizona Parks Department. Rafael, out of Phoenix, was in Yuma for the day. He liked what they showed him as they discussed having the airplane at the Crossing Park. Later, however, the Director of the State Parks Department decided that 56H did not fit into the park's historical theme of "Transportation from 1831 for 100 years."

The committee also held conversations with the Historical Society, Carol Brooks in particular, about how to go about getting grants. She gave them what information she could, including people to contact.

"All of this time, we were going out and I was giving presentations at different organizations," Ron says, "and the media was meeting with us——Greg Gardner at KAWC was interested, and was helping us with the work on the airplane. We had two or three different radio sessions with him. We really didn't do anything on the 48th anniversary, except that I called John Phipps on his radio program that morning and reminded him, and KTTI Radio mentioned it throughout the day. KYMA Television carried a short segment about it, and *The Yuma Daily Sun* had an article about it."

"On November 5th, I went to a city council meeting," Ron continued, "saying that in 1999 we intend to give the airplane to a qualified organization that would have a home for it to be displayed permanently. The council members all voiced their support to find such a place. Joyce Wilson, City Administrator of Yuma, a great supporter of this effort, is continuing to try to find a home, but she also said that it might be two or three years before anything develops."

During the course of all these activities, Judy came up with the idea that the committee consider flying the airplane more than once. It could be used to fly to other airports such as Casa Grande and Williams Gateway airport when they hold their big Experimental Aircraft Association Fly-ins. This would be a way to promote the airplane and the city of Yuma. Bob Bonham, Air Show Coordinator for the Reno Air Races has seen *The City of Yuma* and expressed interest in having it come to the Reno Air Races someday as an added attraction.

"Also, the Historical Society would like to have *The City of Yuma*, if the Jaycee Foundation could come up with some of the funding for a building that could be located behind the Molina Block," Judy explained. Presently, there are old apartments there that need to be torn down. This is part of a long-range plan, consisting of five phases. They are willing to give the committee a plot of land to fabricate a new building in which to put the airplane. The Society had their architect look at it and, based upon the dimensions of the airplane, he came up with a building that would be approximately 6,400 square feet. It would be designed as a museum that would hold more than just this airplane. If the money can be raised somewhere, whether through grants or fund raisers or private donations, that is still a viable option."

"Our first event where *The City of Yuma* was displayed was at the Chili Cook-off downtown on Madison Avenue on the 25th of October, 1997," Ron said. "The airplane, by that time, was pretty well stripped of its paint. We had the engine in it with no top cowling. Garth Worthen made the photo display boards and Shirley Burch made the inserts for all the pictures. The picture boards told the story of the flight and getting the airplane back. Paper airplanes, cut out and folded by Shirley Burch and her Cibola High teacher's assistants and Nancy Woodhouse's T.A.'s, were handed out to all children that attended. These continued to be favorites of young children throughout all occasions when the airplane was displayed. Approximately 4000 of these have been handed out.

A package of information consisting of pictures, video, and documents relating the long history of Yuma's endurance airplane was sent to Continental Engines, Inc., trying to get them as a major sponsor if possible, or to rebuild the existing engine. Initially, there was no response, so another package was sent, finally resulting in a rejection notice. A short time later, Jim Gillaspie and Ed Duppstadt were talking about the situation, and Ed said, "You know, Col. Jim Davis, former Post Commander at Yuma Proving Ground (YPG) retired from the Army and went to work for Continental. In fact, he was one of the Vice Presidents of the Military Vehicles part of Teledyne Continental."

Jim Gillaspie placed a call to Jim Davis, who remembered seeing pictures of the endurance flight when he was at YPG. Jim told him that the airplane was back in Yuma and of the committee's attempt to get Continental to help on the engine. Jim Davis replied that he thought he could help. Although he no longer works for Continental, he is a friend of Carl Bayer, who is CEO of Allegheny Teledyne, Inc., the parent company of Continental. He said that he planned to have lunch with Carl Bayer the following week and would talk to him about the situation. But first, he wanted the committee to develop ideas on how it would benefit Continental, promotionally, to be involved in this effort. He said to send him all the ideas, and he would get Bayer interested. This was accomplished the following day.

Jim Davis discussed the endeavors of the Yuma group with Carl Bayer, who was interested. He thought that the people at Continental might want to participate in the project and said he would discuss it with them. But, as Jim Gillaspie said, "Remember, Continental already said, 'No.'"

Continental's president reconsidered, however, and sent the question to all his department heads, for their opinions. The result of that was a veto by the organization's legal department because of possible liability. The next question from the restoration committee was, "How about parts?" The answer was, "Come to us if you need something."

Later, when the complete inspection of the cylinders had been completed by Jim and verified by Jim Siemens, it was decided that they needed to be overhauled. Jim Gillaspie called DivCo of Tulsa and found out that nobody overhauls that type of cylinder any more because it was considerable cheaper to buy new ones. Ed Duppstadt approached Continental again by way of Jim Davis who in turn talked to Carl Bayer. Nothing was heard for a while, so Ed, who was "leading the charge", called to see what the response was and he talked to the Secretary to the President, who said she would see about getting some help for the group.

A short time later, the group received a letter from Continental saying that they could buy the cylinders at a reduced price ($597 instead of $750 plus shipping) and they would waive the core charge of $150 per cylinder. Continental did not place any special requirements, such as promotional, on the committee.

In November, the airplane was on display at the Colorado River Balloon Festival at Cibola High School. Ron continued to tell *The City of Yuma's* story and the need for a permanent display, which he described as follows:

"In December we contacted Pat Conner, Jim Carruthers, and Bob McLendon and set up a luncheon on the 20th at the Crossing Restaurant. Bob McLendon had a conflicting commitment in Phoenix and couldn't make it, but Pat Conner and Jim Carruthers were there and we discussed the project. We told them what we were doing and what we intended to do, and that we needed a home for the airplane. As a result, a little later, Jim Carruthers mentioned the armory downtown and said that it might be vacated by the National Guard and he thought that the structure was going back to the city of Yuma. He said, 'You know, that would be the perfect place to build a museum.' So that's another possibility."

The first public event of 1998 at which *The City of Yuma* was on display was the Old Time Fiddlers' Contest downtown Yuma in January. The committee had a booth with storyboards and advertising merchandise for sale——tee shirts, caps and mugs.

Judy Griffen Spencer with The City of Yuma

McNeece Brothers Oil Company has generously provided a truck to transport the airplane to numerous events around town, throughout the restoration and promotion process.

On February 7, 1998, 56H was in the Silver Spur Rodeo parade. A work crew moved the airplane from Bill and Barbara Jewett's warehouse to the parking lot at Beeler Equipment Company on Porky Pierson's trailer. There, they mounted the wings and tail feathers back on the fuselage in the dark. The appearance of the airplane in the parade was not widely known in advance. People couldn't believe what they were seeing as it went along the parade route. After the parade ended, the group took the wings off at the south end of Main Street, and moved the fuselage, storyboards, and merchandise to Main and Second Street, where the old Valley Café used to be. Some of these events were not very productive in regard to selling merchandise, but at least word about the airplane was being circulated.

The following weekend, February 13th through 15th, the airplane and merchandise were at the Silver Spur Rodeo. "Yuma Crossing days happened to be the same weekend," Ron said, "and Mary Worthen went to that and took some of the merchandise."

March 6th through 8th, the famous Aeronca was on display at the annual car show, "Midnight at the Oasis," sponsored by the Caballeros de Yuma where they had over 700 cars on display. The City of Yuma had its wings on and cowlings in place.

The committee had thoughts of applying, in 1998, for an historical preservation grant with the state of Arizona. The first step would have been to get *The City of Yuma* into the Registry of Historic Places and Things, for which it would have been eligible in 1998, as it is a 1948 airplane, and the minimum required age is fifty years. When they started, they were told that they had to wait until the historical event was fifty years old, i.e., the flight was in 1949, so no further action was taken.

The committee developed plans for a benefit dinner and auction to be held at Bill and Lois Britain's "Tumbleweed Chuckwagon" in April, as well as a golf tournament.

GRIFF, WOODY, BOB, AND DR. IRWIN AT BRITAIN'S TUMBLEWEED CHUCKWAGON,
APRIL 17, 1998 PHOTO BY PAM SMITH

Nancy Woodhouse led the charge for the dinner and auction and Perry Pensky did that for the golf tournament. Prior to the dinner, Ron and Judy received a phone call from Jacques Istel, who has built "The Center of the World" tourist attraction at the edge of the Sand Dunes west of Yuma. He had heard about the benefit dinner and auction and he wanted to donate the two pilots' names and two others, to be engraved on the wall located there at Felicity. This opportunity would be auctioned off. The offer was accepted and Bob's and Woody's names are now there, engraved on the wall.

"We had interviews on several radio programs during the countdown to the benefit barbecue and the golf tournament on April 17th and 18th respectively," Ron said. John Phipps interviewed Horace Griffen, Jim Gillaspie, and Ron Spencer on his "Talk of Yuma" radio program on KJOK on April 14th, and then he interviewed Horace Griffen, Woody Jongeward, Nancy Woodhouse, and Dallas Hovatter on April 16th. Then in early May, Ron and Jim Gillaspie did a 45-minute taped interview with Kim Sanchez, of KTTI radio. This was aired on KBLU as well as on KTTI.

The dinner and auction at Britain's' drew approximately 340 people and considerable money was raised for the restoration of the airplane. This was a success, largely due to Bill and Lois Britain, Bobby Brooks, Mayor Marilyn Young and the City Council, Nancy Woodhouse, and John Phipps.

The moving crew took the airplane to Britain's' on April 16th, moved it back to the Jewett's warehouse on the 18th, moved it to Tom Pulda's Body Shop on May 1st to be painted, and back again to the warehouse on the 4th. On with the wings and back off with the wings every time it made an appearance! The crew, by this time, had become very proficient at moving the airplane and installing and removing the wings and tail feathers. This crew has varied from one move to the other, but has consisted of: Greg Gardner, Buddy Dean, Garth Worthen, Bill Cox, Ron Zimmerman, Craig Rundle, Jerry Lilley, Walter Aims, Carl Franks, Shirley Burch, Terry Matzner, Jesse Mooneyham, Matt Matlock, Rudy Schantek, Gene LaHaise, and Clay Garrision. Jesse Mooneyham, Therole Miller and Porky Pierson have provided trailers to assist in these moves, as well as McNeece Bros. Oil Co.

Perry Pensky was chairman of the '49 Endurance Classic Golf Tournament at Desert Hills Municipal Golf Course on April 18th. Helping him were Ron Contreras and Kay Nance of Penn Sign Co., Inc. Twelve teams of four took part in the competition and there were 15 tee sponsors. That was a successful venture, providing a good day's activities, and raising additional funds.

Ron gave talks to Kiwanis and Rotary Clubs, and Horace Griffen joined him for a meeting of the Colo-Gila Kiwanis Club. Ron also talked to the Historical Society and the Experimental Aircraft Association (EAA) Yuma Chapter 590. He talked about how the committee needed the EAA expertise and their support for this project. The Chapter, consisting of 30 members, designated Ken Scott as its representative to the committee.

Dan Marries, Bob Woodhouse and Ron Spencer were guest speakers for Louise Kneppers' fourth grade class at Carver School.

Dan showed the KYMA documentary and Bob spent about thirty minutes talking to the kids and answering their questions. The next day the same class went on a field trip to see *The City of Yuma*. Jim and Bob were both there, along with Ron, and answered a lot of interesting and entertaining questions," Ron said.

On May 30th, the Historical Society had a "Rock Around the Block" party in the old McDonald's Department store building in downtown Yuma. It was a fundraiser for them. "We took the fuselage and our souvenir merchandise there," Ron explained. "About 150 to 200 people attended that event. It was a nice party and we had a good time. They had a 1950's theme with old cars, and the airplane fit right in. The Historical Society has been very supportive to the project all along the way."

In June, 1998, the ABC affiliate TV Channel 15 (Yuma Cable Channel 5) in Phoenix aired a segment that was taped in April consisting of an interview with Horace and Jim. The segment was taped for their "Arizona Short Stories" program. They were in town to tape a feature of the Crossing Park. Ray Smucker heard about their coming and suggested they cover *The City Of Yuma* story. Diane Rodriquez, the producer, directed the interviews.

On July 3rd, 56H went to a block party downtown, with its new paint job, logos, and the partially-polished wings in place. Everyone thought that *The City of Yuma* looked "great!"

A web page was established through the efforts of Lyle Blaker, Lowell Vahl, and Arizona State University. It contains information about the flight, the restoration of the airplane, and the search for sponsors and participants. The address is

http://aztec.asu.edu/enduro49/.

"On September 12, 1998, we wrote to Col. Turner at the Marine Corps Air Station," Ron says, "asking for their support and help for the observance of our fiftieth anniversary, on October 10, 1999. We were asking to be able to use the airfield, and hoped that the military would have various static displays of airplanes and equipment."

On October 10, 1998, the 49th anniversary was observed at Bet-Ko-Air, with 150 to 200 people attending. Jim Gillaspie, Ron, Judy, Martha Taylor, Shirley Burch and Nancy Woodhouse all helped to plan the event. (Jim was able to escape to Albuquerque, along with Bill Jewett, each with a balloon, for the huge annual Albuquerque International Balloon Fiesta, so they were flying high for ten days.) The Yuma High Choralairs performed at the anniversary observance, in their usual enthusiastic and excellent way. The Crossing Restaurant catered the food, and Dan Marries of KYMA was the master of ceremonies. He showed his excellent documentary video about the endurance flight. Ray Smucker was there from his home in Phoenix, with his wife, Nadine, to bring back memories for some and to tell others how the flight came about. David Schuman was presented with a commemorative bomber jacket and Ticket Number One for a flight in *The City of Yuma* as he attained "Crew Member Status" with a generous donation to the restoration fund. The committee formally thanked Martha Taylor and her crew at Bet-Ko Air, for providing the hangar and setting the event up. Sponsors of the event were: Coca Cola, Crystal Waters, Johnny's TV who provided a big-screen television set on which the documentary video was shown, Fisher Chevrolet, Plaza Auto Center, Sun Valley Beverage, Yuma Crossing Restaurant, Todd Pinnt and the Civil Air Patrol cadets, Jaycee Foundation members, Jaycees and the Junior Jaycees. It takes a lot of people to conduct such an event.

At the end of October, 56H and the commemorative merchandise went downtown for the Historical Society's Chili Cook-off. Many people commented about the dramatic change that was in evidence after a year of restoration work.

On November 21st and 22nd, *The City of Yuma* was at the annual Colorado River Balloon Festival at Cibola High School. Helping with the appearance were Claude and Jo Beth Sharpensteen, the Civil Air Patrol Cadets, Nancy Woodhouse, Theresa Buscko, Laura Pully, Judy Spencer, and Elton Worthen.

Ron says, "On January 12th, we began to plan an event to observe the fiftieth anniversary of the landing of *The City of Yuma*, "We decided to try to have a function at Paradise Casino on the night of the October 8th, with entertainment representing the 40's or 50's era. Larry Boyd volunteered to have us come to his farm in the Dome Valley for a breakfast on Saturday morning, the 9th. Then we wanted to have a lunch at Somerton. We talked to Stan Lawless about that and he said 'OK!' That evening the Jaycees will host a barbecue dinner, cooked by "Chef" Elton Worthen, at Bet-Ko-Air.

THE MOVING CREW PUTTING ON THE LEFT WING FOR A PUBLIC APPEARANCE OF THE CITY OF YUMA.

Ron, Jim, and Horace gave a presentation at Cocopah Bend R.V. Park on January 18th. Then the airplane was moved to KYMA at the end of January to help them observe their eleventh anniversary.

On February 2, 1999, Jim and Ron took their "wish list"— for Sunday afternoon, October 10, 1999, to Lt. Col. Coburn, the Operations Officer at MCAS—things that might happen at the observance of the fiftieth anniversary. These included a static display with modern and vintage military aircraft if possible, civilian vintage airplanes, possibly some agricultural airplanes, and an AV-8 (Harrier) flying demonstration. Then *The City of Yuma* would fly by at a low altitude with the 1948 Buick refueling car running along under it, to simulate a refueling run. A military air escort would be the committee's first choice, or a civilian aircraft escort. Lt. Col. Coburn said that Col. Turner, commander of the base wants to see this fiftieth anniversary event happen and so does he. They have to obtain official approval from MAWTS, the Military Air Weapons Tactics Squadron because, at that time, their Weapons Tactics Training will be going on and they have to be sure that there is no Sunday afternoon air activity.

Ron says that Lt. Col. Coburn felt, "I feel that we can work around that."

"We asked them if they could take care of tents, seating, and security. Our merchandise would be there for sale, and we would have a dedication of the airplane to the city of Yuma—probably to Mayor Marilyn Young."

Then on the 12th through the 14th, the airplane was on display at the rodeo, along with the souvenir merchandise, followed closely by the big annual Air Show at MCAS for "Military Appreciation Days," on the 20th.

RON SPENCER, GENE LAHAISE, CLAY GARRISON, AND JIM GILLASPIE WITH THE CITY OF YUMA ON DISPLAY AT THE JAYCEE'S 1999 SILVER SPUR RODEO

Transporting both wings for a public appearance. Clay Garrison, Carl Franks, Jim Gillaspie, and Ron Spencer facing camera, Gene LaHaise on near side

McNeece Brothers Oil Company had offered a stake bed truck to transport the airplane to the air show, but its engine had burned up. Garth Worthen arranged for a truck to take the airplane out to the airport, and Jesse Mooneyham came to the rescue with a trailer to take it back to the Jewett's warehouse afterward.

The air show was the first event that featured a newly-acquired 1948 Buick Super Convertible along with the airplane. George Murdock had been looking for a car like the first new one that he had bought as a young man. He remembers that he paid $2,800.00 for that new Buick convertible. Milton Phillips, regional authority on old cars, had helped him search for one over the years, and finally located this one in Carmel, California. Milton would like to have bought the car himself, but he knew how much George wanted one, and he drove out to George and Shirley's home in Roll and told them about it. Milton made the deal in a phone call to Carmel within minutes. George had bought the car, sight unseen, on the basis of the description by his friend, Milton, who then arranged to have the car shipped to Yuma. The fifty-one-year-old

Buick is in excellent condition, with its original dark green paint, but as George says, "It needs to be white," as his original car (the refueling car) was. It is being restored, but not at the expense of the Airplane Restoration committee. This is George Murdock's responsibility. He has loaned the car to the committee for the events since its arrival in Yuma and he plans to do so for the observance of the fiftieth anniversary. Tom Pulda's Body Shop and Jimmy Allen are preparing the car for Jimmy to paint. Dave Garcia at Dave's Auto Glass, is happily restoring the interior of the car with red leather seats and red carpeting.

The Buick convertible attracted a lot of attention along with 56H at the air show, and the same thing happened at the car show, "Midnight at the Oasis." The beautifully- restored airplane and the dark green convertible were displayed together on the expansive lawn at the Caballero Park and attracted a host of admirers.

THE CITY OF YUMA WITH THE 1948 BUICK SUPER CONVERTIBLE, BELONGING TO GEORGE AND SHIRLEY MURDOCK, SIMILAR TO THE ONE USED IN 1949. AT THE MIDNIGHT AT THE OASIS CAR SHOW, MARCH 5-7, 1999

The various commemorative items were sold and donations were gratefully accepted.

In publicizing the dinner and auction on March 13th, Bob Woodhouse and Ron were interviewed by John Phipps on his "Talk of Yuma" program on KJOK, as well as by R. J. Bones at KTTI radio, and Dan Marries at KYMA.

"At our dinner and auction," Ron says, we had a good time and we made a profit at both events. The Fort Yuma Rotary Club solicited items for the auction and took care of the ticket sales. John Phipps and Christy Jo Johns did the auction for us." John and Christy Jo both have the auctioneer's language mastered and that was entertaining and effective. Bobby Brooks cooked delicious tri-tips and Mayor Marilyn Young and other members of the City Council served the food. The Civil Air Patrol cadets were there in their uniforms. It was a special occasion.

LOADING THE CITY OF YUMA TO BE TRANSPORTED BACK TO THE JEWETTS' WAREHOUSE, SUNDAY MORNING, MARCH 14, 1999, AFTER BENEFIT DINNER AND AUCTION AT BRITAIN'S TUMBLEWEED CHUCKWAGON

The City of Yuma on McNeece Brothers Oil Co. truck for transport across town, March 14, 1999.

No small item is the matter of keeping track of the finances along the way—— income and expenditures, bank account business, financial statements, and treasurer's reports at the monthly meetings——Mary Worthen is treasurer of the committee, and the other members appreciate the time that she spends on that important business.

"On March 26th, we received a check from TOSCO Corporation and Union Oil in the amount of $2500.00," Ron says. "We had a picture taken for *The Yuma Daily Sun*, with Mackie Gill of Sellers Petroleum presenting the check."

On April 14th Erik Bowan and R. J. Bones of KTTI had decided to put on another endurance event, staying in the airplane for 47 hours and attracting attention, through their broadcasting, to try to raise another $5000. May 6th started the "47-hour Endurance Day" at 4:00 P.M. and their plan was to stay in the airplane until 3:00 P.M. on the 8th. When they were just 15 minutes shy of their "escape time," they were still short of their target amount of $5,000. Bill

Ruch from the House of Vacuums, and David Schuman from Farmers Insurance drove into the parking lot and wrote the checks that were needed to bring the total to their $5,000 goal.

LOADING THE AIRPLANE ONTO THE LOWBOY FOR THE START OF A 47-HOUR "ENDURANCE SIT-IN" BY KTTI DISC JOCKEYS ERIK BOWAN AND R. J. BONES

THE CITY OF YUMA DURING "47-HOUR ENDURANCE SIT-IN" BY ERIK BOWAN AND R. J. BONES OF KTTI. MAY 6-8, 1999

"On the 18th, we met with Paradise Casino people and they gave us some good suggestions, and soon after that they obtained the approval from the Tribal Council, so that they can conduct a big event on the evening of October 8th .

During the years of the existence of the Yuma Municipal Airport building, a hand-painted mural on the wall showed the mistaken information that *The City of Yuma* flew 1008 hours, rather than the correct figure of 1124 hours. That error carried over into the new terminal and had to be corrected later. This was done and the World Endurance Flight record of 1949 is now shown correctly.

Based on a meeting held on June 24, 1999, the 50th anniversary activities will be held at Bet-Ko-Air in lieu of the Marine Base or the new terminal. Present at that meeting were: Lucy Shipp, Yuma County Board of Supervisors; Lt. Col. Coburn and Cpt. Jimenez, MCAS Yuma; Ed Thurman, David Gaines, and Daren Griffin, Yuma County Airport Authority; Chauncy Dunstan and Mike Covey, Caballeros de Yuma; Martha Taylor, Bet-Ko-Air; Ken Scott, Chapter 590, EAA; and Dustin Dinwiddie, CAP Cadet Squadron 509, Yuma.

The Caballeros de Yuma have made a commitment to help with the Fiftieth Anniversary Celebration on October 8th, 9th, and 10th, 1999. Chauncy Dunstan is their delegate to work with Ron and Judy Spencer and their group.

Several people have donated a few thousand hours each and many have given hundreds of hours in the restoration and promotion of *The City of Yuma*. The original flight required the time and effort of a large number of volunteers for less than a year and this present-day effort has taken a smaller number of people a much longer period of time.

Chapter Five

Old Timers Remember

Through the years, people of Yuma have kept the 1949 Endurance Flight of *The City of Yuma* alive and well in their hearts. It was especially so for Horace Griffen. It has probably been one of his most favorite subjects of conversation, and he has kept in touch with just about everyone who was involved in the flight.

EVENING REFUELING RUN SHOWING WOODY BEING CONGRATULATED BY A VISITING DIGNITARY.
PHOTO BY EMIL EGER

Several of the people who were most involved in the flight have passed away and are sorely missed by those who remain, especially as the fiftieth anniversary approaches.

Paul Burch was killed in a tragic accident at the Yuma County Airport in December of 1979. The pilot of a military jet, an A-4, ejected on takeoff when the airplane lost directional control. He didn't shut the engine down and the aircraft continued, under full power and unattended. It was headed northeast, then veered about 30 degrees toward the north and crashed into the hangar where Paul was in his shop. He was killed instantly and the pilot was found walking on the runway. Paul Burch was a prominent figure in aviation history in Yuma. In addition to being an aircraft mechanic at Marsh Airport for many years and then having his own business, Burch Aviation, he was a major founder and the first president of the Airport Authority. He was often referred to as "a super mechanic."

Charlie Gilpin's death in December, 1985, was a shock——sudden and unexpected. The cause was toxemia resulting from a gall bladder infection. He had retired as General Manager of Gilpin's Construction Co., Inc. in 1977, turning the business over to his daughter Jeffie's husband, Don Riley. Charlie did this in order to be able to spend more time with his wife, Julie, who had contracted cancer the previous year. Julie passed away on January 1, 1978. Charlie continued to own the business and remained Chairman of the Board, but didn't work there. He married Ann and they traveled extensively for several years before his untimely death.

Bernie Pensky died of cancer in July of 1990. He and Woody Jongeward were next-door neighbors in their businesses, and Bernie's son, Perry says that Woody probably asked Bernie about having Penn Signs paint all the signs on *The City of Yuma* and on the refueling car, and maybe even about serving on the morning refueling crew. Griff refers to Bernie as being "the tall one on the morning crew." Bernie and Bob Hodge were good friends, dating from their military pilot days at the air base. Perry mentioned that a lot of Yuma businessmen had been stationed at the air base and came back to Yuma after they got out of the service, because they liked the town so well.

Perry also said that, when he was young and into sand rails and motorcycles, Bernie worried because those things were "dangerous." Perry says, "I said to him, 'When you were my age, you were standing up in a speeding convertible, handing cans full of gasoline to some guys in an airplane that was right there beside you, and you think what I'm doing is dangerous?'"

Bob Hodge also died of cancer, in January of 1995. He went to Scripps in La Jolla in 1994 and was diagnosed with "a very fast-acting terminal cancer in the top of his head and in his lungs," his son, Mike said recently. He went to Tucson for a second opinion and was told the same thing. The prognosis was that he probably only had four to six months to live. "He didn't whimper on the way out," Mike said. "He was very courageous and never lost track of who he was."

Mike said that, when his father became frail, he didn't want other people to see him. He wanted them to remember him the way he had been before. Mike also said that Yuma has changed so rapidly, and that most people don't remember those old timers like Bob Woodhouse, Woody Jongeward, Charlie Gilpin and Bernie Pensky. "They were pioneers, explorers, adventurers; they tested the envelopes in their own ways."

There are others, too numerous to mention, but individually missed by those who worked and played with them.

The endurance flight of 1949 was one of the most exciting and interesting things ever to happen in this corner of Arizona, and it had a major effect on the economy and the fortunes of those who settled and invested in the area. A sudden bright idea, then grasped, pursued, and effectively carried out, made a difference.

Old timers enjoy reminiscing about the days when *The City of Yuma* held their attention and Bob and Woody were the heroes of the day. The spirit of the community is a thing to be remembered with some sentiment and appreciation. The high point of the day for everybody was the refueling, including the many citizens not directly responsible for any aspect of the flight. The exchange of gas, oil, and food between the car and the airplane never failed to excite the expectant crowds at the airport. The evening refuelings became the place to be. The press always reminded Yumans that they were welcome, but that they must stay clear of the runways.

The early-morning runs were not as popular as a spectator sport but the wives, Berta and Betty, were out there for every refueling run, morning as well as evening. Griff says, "There wasn't one run made, during the entire flight, without either Berta or Betty"—— and they both had full-time jobs. One morning in the early stages of the flight, the wives started handing fuel cans up to the pilots when someone didn't show up on time. After that they were "officially adopted" by the morning refueling crew and they took an active part. Sometimes they were boosted up by a couple of members of the ground crew to within "smooching distance" of their husbands.

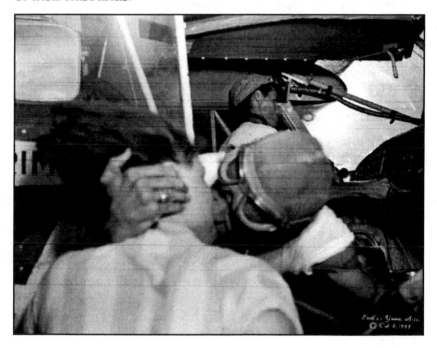

October 6, 1949. Bob and Berta sharing a high-speed kiss the day after the existing world's record of 1008 hours was broken. Photo by Emil Eger.

Pictures of that activity made front pages of newspapers and some newsreels. In a letter handed down to his parents, Bob wrote, of the wives: "They are always out there and all smiles, even at the early-morning refueling. I think it's harder on them than it is on us. We're getting more sleep than they are, by far."

There was an incident that livened things up a bit during one of the refueling runs and Bob told how it happened. "On one occasion, 'Big Bob Hodge,' who was 6ft.7in. tall and weighed about 275 pounds, was in the refueling car and somebody else was handing us a gas can. Well, Hodge reached up and grabbed the strut to try to stabilize the airplane. We couldn't really tell where that force was coming from, but all of a sudden the airplane was tilting to the right. So what did I do? Well, I gave it a little more left aileron. And what did Hodge do? He pulled it down more. I'm thinkin' that the airplane is going haywire somehow. Well, he finally got all of it he needed and he turned it loose. So then the airplane went 'Whoom!' with the right wing high. Ol' Hodge was probably holdin' a hundred pounds of weight down and I'd been tryin' to overcome it." Several people have mentioned that Bob Hodge lifted those gas cans up as easily as if they were empty, "never mind the wind resistance as the car and airplane sped down the runway."

Probably the question most often asked is "How did you handle your bathroom routine?" Bob says, "Well, Berta went down to Imperial Hardware and bought an aluminum pot with handles on each side. Somebody came up with some insulated bags that were made out of two layers of a kind of asphalt-type stuff with chopped-up paper in between the layers. At that time, you couldn't just go buy plastic bags. These were waterproof and insulated and they had a little wire or something that you twisted around the top. That bag would just fit in the pot and the edges would turn down a little ways, so whenever you got whatever you wanted in that pot, then you'd pull it up and twist that little wire around the top and then we'd fly over to California and throw it out, 'cause we had heard that they needed the water over there."

AIR-TO-AIR SHOT SHOWING BOB DRYING OFF AFTER TAKING A SPONGE BATH. PHOTO BY ROD DALEY

One day Griff told Bob, over the radio, that a load of new Buicks had come in, and he said, "You guys ought to fly over and take a look at them and pick one out to buy when you land." Bob said, "Well, bring them out to the airport for us to take a look at and we'll make a deposit on one of those suckers!"

The Yuma Daily Sun had an amusing article on September 13, 1949. "Today marked the 20th day in the air for endurance fliers Bob Woodhouse and Woody Jongeward and also instituted 'The Big Shave.' True to his word, state Junior Chamber of Commerce president Ray Smucker fulfilled his promise this morning to rid his scalp of its hairy covering. Sixteen men and five lariat ropes finally subdued the Herculean head of radio station KYUM, so that Cecil Huling of the Sunshine Barber Shop could do the honors. Following in the footsteps of his idol, Sampson, Smucker became meek as a kitten as soon as his locks dissolved partnership with his noggin.

"Getting back a little closer to what actually happened, Woodhouse and Jongeward, after a slight conference of two weeks on the subject, sent down word that they would settle for a half inch of hair to be left. As soon as the decision was handed down, a loud cheer went up from the ranks of the Jaycees on the ground.

"This morning, as usual, Smucker was making his regular morning blowcast when the door flew open and in walked Yuma Jaycee prexy Charlie Gilpin and Barber Cecil Huling. Without so much as a goodbye phone call to his wife, Smucker allowed the towel and apron to be fastened in place, the clippers plugged in, and his tresses removed while he continued his work at the microphone.

"Each day hereafter, Woodhouse and Jongeward will fill out one 'Big Shave' ticket which entitles the person whose name is enclosed to one free skinning at the Sunshine Barber Shop. The bearer must have the ticket honored by 10 A.M. All recipients of "The Big Shave" tickets who take their turns peaceably will be allowed a half inch of hair. God help the stubborn. ....Who will be Number Two? Read the next exciting episode in tomorrow's *Sun*."

Sometimes the playful refueling crew had a surprise dreamed up for the pilots. Suggestive reading materials were not unheard of, and on one occasion the pilots' gift of the day was a new exercise pad made up of several dozen realistic-looking "falsies" of all sizes and shapes.

Another time, according to *The Yuma Daily Sun*, a well-meaning refueling crew member included two beers in an ice chest with the usual orange juice and milk that was handed up to the pilots. They couldn't drink the beer, so Woody opened both cans and "returned the brew to the refueling crew on the next pass."

The pilots devised ways in which to amuse themselves. Bob described one of their little schemes that brought a reaction from a major sponsor. "Sometimes we'd fly over the dump which was down by the river in those days, and we'd pick out a target to bomb with oil cans. But then the guys that were giving us the oil, Union Oil Company, decided that all of a sudden the oil consumption on the airplane had gone up considerably. We were using two or three quarts more each day that we had been using. Well, we told 'em, 'You know, with empty oil cans, you throw one of them out and it just goes pffffft—and you can't hit anything with 'em, so we're using full ones.' So they sent up some grapefruit for us to do our bombing with."

On another occasion, Bob and Woody told about an amusing little trick that they played on one of the refueling crews. The fliers asked the morning refueling crew to help them conduct their ruse. They sent a note down, asking the group to go to the junk yard and find the nastiest, greasiest old carburetor that they could, and put it in a cotton seed sack. The men were able to get an old carburetor from an airplane, and they passed it up to the pilots the next morning. During the evening refueling, they said to the crew: "We've got something here that we have to get fixed right away, so take this to Paul and get it fixed as soon as possible," and they handed them the old wrapped-up carburetor.

Some people remember that Woody asked for a first-aid kit one morning, "including some snake-bite medicine." He said that Bob had been bitten by a rattlesnake. The truth was that Bob had a recurring dream in which he came upon a coiled rattlesnake and he kept trying to get away from it. His family members knew what it was like to be near him when he was having a nightmare, and all felt sympathy for Woody in that three-foot-wide cabin with him. There was also a question of the real need for "snake-bite medicine."

The cook at the Valley Café during those years was Henry Casares and he still lives in Yuma. He remembers seeing *The City of Yuma* flying over all the time and the excitement about it. He knew just about everybody in town, as the Valley Café was "the place to see and be seen." Henry worked the 5 P.M. to 1 A.M. shift and remembers seeing the airplane flying around Yuma in the early morning hours. He stated that he showed his daughters the airplane and told them that they were "looking at history and that is when things became meaningful." He offered as an analogy that you can read about the Grand Canyon all you want, but when you actually go look at it and it takes your breath away, you truly understand what a marvel it is.

Then there was the time that the pilots "dropped the watch," or, more accurately, "the watches." Everett Self owned The Sweet Shop in Self's Enterprise Building in Somerton. He had a soda fountain, a news stand, a bus station, and even a jewelry department and watch repair business. That was "the in place" in Somerton. Charles Phillips was the watch repair man. Everett and Charles asked if the pilots would drop two Wyler shockproof watches from an altitude of 1000 feet, onto the concrete runway at the airport, to prove and advertise that they were shockproof. The plan was agreed upon and Charlie Gilpin placed an empty gas can on the runway as a target. He told the pilots that, if they came within six feet of the can, the watches were for the refueling crew. Paul Dawson of KYUM helped with the publicity, but he had been dubious about the claim and asked the editor of *The Somerton Star* newspaper to bring along a broom and dustpan to sweep up the pieces. About 500 spectators attended a refueling when the watches were dropped with red, white, and blue streamers attached to them with adhesive tape. The whisk broom and dustpan were not needed, as the second hands on both watches were moving along normally. There were several "official judges" of the condition of the watches, which were also inspected by about 50 spectators. Charlie Gilpin asked the pilots whether they wanted the watches that they had dropped or new ones. Everett said, "They enthusiastically voted on those that they had helped to prove were shockproof."

Everett had taken 16mm movie film of the refuelings and other activities in 1949 and he had that footage transferred to video for the fortieth anniversary dinner at Chretin's. He had copies for sale at that time, and he received a round of applause from the assemblage for that.

Bob Hodge and Jeffie Gilpin Riley organized the celebration of the fortieth anniversary of that big day in 1949, at Horace Griffen's request. Griff served as Master of Ceremonies at a Rotary meeting at lunchtime on October 10, 1989, and then at the dinner at Chretin's in the evening. He introduced all of those in attendance and when he came to Bob and Pauline Hodge, he said, "Bob was chosen for the refueling crew for an obvious reason. In those days he was taller than he was wide."

Later, when people were telling stories of the days of the flight, Bob Hodge said, "One of my favorite stories is about Bobbie and Woody. They were up in the air and we had a hamburger spot downtown and there was a girl down there that was built like you- know-what——Genola Gray. So Bertie and Betty were in the convertible with me and we went down to get a hamburger, but the boys didn't know that the girls were in the car. They said, on the radio, 'Where you goin'?' and I said, 'We're going down to get a hamburger.' They said, 'Oh is Genola workin'?' I said, 'I don't know, I'm gonna find out.' So we got there and they said, 'Genola?' So I asked her, 'Do you want to speak to the boys?' and she said, 'Sure. Hi, Bobbie! Hi, Woody!' 'Oh, Hi, Genola!' So they chatted back and forth and Bertie and Betty and I were eatin' our food and we got all through and the boys said that they wanted to speak to Genola again, so they said, 'Genola, when we get back down, the first thing we're gonna do is come down and get a hamburger and a malted milk.' I said, 'Is that right? The very first thing?' They said, 'Yeah!' I said, 'Well, I've gotta leave you guys, but there's one more thing I want to do,' and I handed the microphone to Betty and she said, 'Good night, Woody,' and she handed the mike to Bertie and she said, 'Good night Bob!'——
——and there was total silence!"

The airplane droned around Yuma, ventured into California, and paid Phoenix a visit on the occasion of a parade honoring Jacque Mercer who brought the Miss America crown home to Arizona and to her home town of Phoenix. The endurance airplane flew over the parade and then circled the state capitol building as Miss America received the official greetings of her home state.

On the 22nd day of the flight, the pilots acted as messenger boys for the Yuma and Phoenix Rotary Clubs, according to *The Sun*. Rotary was sponsoring appearances of Horace Heidt with his show in both cities. When the Phoenix organization ran short of publicity posters advertising the Horace Heidt Show, the fliers were approached in regard to running an errand for the Phoenix Rotarians. They complied and were scheduled to head their light plane in the direction of the capital city today just as soon as permission had been obtained for them to drop the bundles at Sky

Harbor Airport. An excellent photo captured the moment when the package was dropped.

THE CITY OF YUMA DELIVERING A PACKAGE FOR THE HORACE HEIDT SHOW IN PHOENIX, SEPTEMBER 15, 1949

That was the same day that a 4:00 A.M. refueling was added to the schedule, as described in the second chapter.

The pilots' travels included four trips to Phoenix and, on other days, they flew to Tempe, or Chandler, Winslow, Coolidge, Tucson, or Casa Grande. An article in *The Yuma Daily Sun* indicated that one of their favorite tours was to Los Angeles and along the coast from there to San Diego, "checking out the beaches."

On one occasion the pilots suddenly found that they had flown far away from their usual territory. Bob related the story at a modern-day meeting.

"One night Woody woke me up and he was lookin' around——kinda scared lookin' ——and he said, 'Bob! What are those lights there?'

"I said, 'I don't know. Why?'

"He said, 'Well, I've been asleep!'

"And you could sleep in that airplane. I had gone to sleep when I was flyin' and he'd gone to sleep before. He might have slept quite a while that time, but if you flew maybe three or four minutes, the airplane would usually begin to drift off one way or the other and the engine would speed up, 'cause, you know, it had a fixed-pitch propeller, so the sound would change and you'd wake up. So, I'm lookin' around and I'm thinkin'—you know, I can't see Mexicali and I can't see Yuma.

"I said, 'Well, whatever town that is, it's back behind us, so I'd suggest we turn around and go back there and see what it is.'

"Well, it took us almost an hour gettin' back there. We figured we were the other side of Puertecitos, maybe Gonzaga." Both of those are fishing villages on the Gulf of Lower California, some distance "south of the border."

A Yuma veteran was a patient at Fort Whipple Veterans' Hospital in Prescott. He knew that the pilots traveled around and he wondered if the fliers might come and fly over the hospital so that he could see them. They honored that request, flying over the hospital at an altitude of about 1500 feet on September 28th. That was the day that the pilots had completed five weeks in the air. With only another week to go before breaking the record, the local news was full of plans for a "reception" as the pilots flew low over the runway. Yumans were advised just where to line up their cars and the suggestion from the Jaycees was that each person take a white tea towel or something white to wave at the pilots. The Yuma Bus Company provided free rides to and from the airport for the refueling that evening and on other occasions. Yumans were advised that they could stop the bus at any point by waving a white tea towel. Bob and Woody were surprised to see a crowd gathered at the airport, estimated at 5000 people, whereupon they received their salute of the white tea towels. Such demonstrations gave them the needed boost for the long hours of living in a small space, approximately 6 feet x 3 feet x 4 1/2 feet, for almost seven weeks.

Another highlight of that event was the deputizing of the fliers by Sheriff Jack Beard. The official cards were sent up to the pilots during the evening refueling. Also two new blankets were handed up to the pilots, donated by Horace Porter of "The United," a store in downtown Yuma.

The Yuma Daily Sun article on that day stated: "Folks attending the refueling will probably notice a change in the title of the make of the pickup which is used for hauling the gas cans and oil to the airport. On the tailgate of the truck, the title did read "Dodge," but now reads "Hodge." The truck and the fuel to keep the very important vehicle going were donated by Bob Hodge, General Petroleum distributor in Yuma. Hodge is also on the board of directors of the Yuma Junior Chamber of Commerce and a member of the evening refueling crew." (Ed. note: At one time the truck for that purpose was provided by Union Oil Company, as noted elsewhere. Apparently that responsibility changed at some point during the flight.)

On September 22nd, the 29th day of the flight, the Jaycees sold coffee and sandwiches at the airport. The Swift Company exhibited a six-foot-long hot dog in the window of the L & R Market that day and said that they would make sandwiches from the meat in that display, and sell them at the airport that evening. They also made coffee and sold it, donating the proceeds to the Jaycees who used the funds to help defray expenses associated with the flight. The sandwiches sold for ten cents and the coffee for a nickel! The article in The Sun hinted, "There just might be some sort of added entertainment thrown in on the deal, for attending flight boosters."

At one point, during the second attempt, The Yuma Daily Sun carried an account headed: "Endurance Fliers Hear Jaycee Song. Never to be outdone on any count, the Yuma Junior Chamber of Commerce is building up more momentum every day in putting over its endurance flight. When pilots Bob Woodhouse and Woody Jongeward brought their light ship down from the blue this morning to refuel, they were greeted over the radio with a song. While the fliers circled the field prior to swooping down to pick up fuel and supplies, the ground-crewmen sprang the latest Jaycee surprise——a song, composed principally by Bob Hodge. The words are sung to the tune of 'McNamara's Band.'"

Here are the words:

'Oh we've got a couple of pilots of whom we're mighty proud.

We're going to shout their praises and we'll shout them clear and loud.

Their names are Woodhouse and Jongeward and the two of them can't be beat,

And they'll be leading the whole parade when they turn on the heat.

Give 'em a ship that they can fly, Now watch these boys, they're mighty sly;

If the ship don't quit, it's do or die and we'll watch the old endurance record go bye bye."

The article continued: "The song was born only this morning, and at noon several hundred copies were being circulated around town. Jaycees predict that people all over the state and even the country will be hearing their renditions shortly."

Another time the ground crew wrote a little song for them, and sang it to the tune of "There Are Smiles That Make Us Happy."

"There are guys that make you happy,

There are guys that make you blue,

But at 5:00 AM this morning,

Those two guys looked just like you.

There are days when we are happy,

There are days when we are blue,

But the days will all be happy

As long as you're in the blue."

The pilots answered in kind:

"There are men that make us glad,

There are men that make us blue,

There are men that we love most to see,

the men on the refueling crew.

There are men like Gilpin, Hodge and Michaels

There are men like Louie Mueller, too,

But Oh, men, you fill our tank with gas

And that's why we go for you."

Still another about the refueling, to the tune of "Honey, Baby Mine:"

"Here comes a man with a can in his hand,

He's got gasoline in that can.......Honey, Baby mine.

Gasoline in that can is yours,

Hurry, hurry open your doors......Honey, Baby mine.

Lean way out, reach way down,

Be damn sure you don't touch the ground....Honey, Babymine.

Hurry, hurry, get in a rush

We're almost clear down to the brush.......Honey, Baby mine.

Jongeward, hurry and get that stuff.

We're not taking any of your guff......Honey, Baby mine."

An article in *The Yuma Daily Sun* on September 6, 1949 states: "Another song creation was born over the weekend by Shirley Woodhouse, sister to pilot Bob. Using the music of 'Ghost Riders in the Sky,' Miss Woodhouse concocted 'Endurance Fliers in the Sky.' The fliers were introduced to the song Monday morning by the refueling crew. It will probably reach the local public before long." But the words to that song seem to be lost to history.

And a poem, written by Lucille Foos:

"High over Yuma by day and night
Are happening many things.
And every Yuman stands and cheers
At a flash of silver wings.

Why does everyone stand with bated breath
And maybe breathe a prayer?
It's something great, haven't you heard?
Yuma is in the air.

Two men of Yuma have gone aloft
To prove by actual test,
That no matter how good flying weather is,
In Yuma it's always best.

So ride on, Knights of the air,
May your engine never fail,
But waft you gently through the sky
'Til you hit the homeward trail.

You're not alone as you wing your way
Up there in the Heavenly blue,
But every Yuman's thoughts and prayers
Are riding there with you.

Think of California's face when you've won the race

As you take your needed rest,

Bring that record down to your home town,

Yuma, the star of the West."

Ironically, just a few minutes before that record was broken, someone landed at the airport in an Aeronca Sedan——exactly the same kind of airplane as *The City of Yuma*. It was dark enough that the absence of signs and slogans wasn't easily seen. A few Yumans almost had heart failure. That pilot could hardly believe the sensation that he created.

After the flight, interest continued to be evident around the country. The endurance fliers flew *The City of Yuma* to Chicago where they rode a float with the airplane along Michigan Avenue for the National Jaycee's convention in 1950. Betty and Berta said, "Not without us!" and they accompanied their husbands on that trip.

Bob's parents, Harold and Ethelind Woodhouse and his sister, Shirley, made that same trip, flying along beside the endurance fliers and their wives in another airplane.

A former neighbor, George Foelsch, had moved from Roll to his home state of Wisconsin and he said to the Woodhouses, "If you'll fly my Stinson Voyager from Roll, back here to Wisconsin, I'll pay your airlines fares back to Arizona." Shirley was just home from college in Colorado and she was delighted to go along, to visit relatives in Wisconsin and then attend the Jaycee Convention in Chicago and see *The City of Yuma* and the pilots on a float in a big-city parade.

As the Arizonans were flying over Kansas, Betty Jongeward said, "Boy! If you were going to buy a farm in this part of the country, you'd have to buy a used one, wouldn't you?" She wasn't accustomed to seeing every acre of land under cultivation.

A model airplane was built as an exact replica of *The City of Yuma* with all the same signs and slogans, but with a wingspan of six feet. It was one-sixth the size of the original record-holding Aeronca.

MODEL AIRPLANE, THE CITY OF YUMA, JR. (1/6 SCALE) BUILT BY BOB McFARLANE

Yumans Thurman Hart and his son Billy were the "co-pilots" who flew the model on a 100-foot line, controlled from a jeep driven by Herschel Wright. The Jaycees sponsored a second endurance flight, with the model which they called *The City of Yuma, Jr.*, setting a record for model airplanes, on October 10, 1952, the third anniversary of the record set by its "big brother."

Young "Billy" Hart grew up and is now the Principal of Taft Elementary School in Santa Ana, California. He sent to the authors an amusing saga of the endurance flight of *The City of Yuma, Jr.*, as follows:

"For what it's worth —— a little more history on the bird.

"Bob MacFarlane built the airplane and then recruited Dad and me to fly it for him. I was the Arizona junior state model champion at the time and was supposed to know how to fly the thing! We test flew at Marsh Airport, down by the river, and promptly tore the bird up! Dad helped rebuild the thing, removing the wing warp and bringing the balance way forward. He also incorporated two fuel tanks in the wings in order to keep the weight on the center of gravity.

"A model on control lines flies in a continual circle with the pilot, who controls pitch only, standing in the center. Following the repairs, the bird flew like a dream, but was fast. She flew at 60 to 70 miles an hour (we had no throttle, of course!) and even on 100-foot control lines the bird would be impossible to catch and refuel in mid air.

"I think it was Paul Burch, (super-mechanic at Marsh and second dad to me) who came up with the idea of flying the bird from a moving jeep.

Model Airplane, The City of Yuma, Jr. during its own endurance flight. October 10, 1952.

We could travel in almost a straight line, giving our brave refueling crew in a pickup the opportunity to catch the plane and feed it.

"We flew several practice flights at the county airport and things went well. On the morning of October 10, 1952, we took off to try for the record. I was given the honor of being the first pilot and I remember the sudden feeling of fear in trying to get the bird off the ground. It seems that we never made a practice flight with full fuel tanks, and the bird weighed a ton! We lumbered into the air and the resulting pull on the control lines about took my young arm off! It was definitely time for the second pilot (Dad) to fly this thing!

"Russ Phillips solved the problem of heavy pull on the control lines by driving the jeep in as straight a line as possible, making a very wide 180-degree turn and traveling back. Also, as the fuel burned off, we became lighter and it was easier to fly.

"Refueling was another matter. I was flying when Dad decided it was time to feed the bird. Russ drove as straight as possible and the wild-eyed crew in the pickup came thundering after us — in one of the flight photos you can see a long plastic tube trailing from the bird. The plan was for the guys in the pickup to grab the end of the tube, push it on a fitting screwed into a two-gallon tank of gas-and-oil mixture, and then pump like crazy on a tire pump to pressurize the tank, which would move the fuel up the line and into the tanks. When finished, they were supposed to pull the plastic tube off the pressurized tank and we were free to go....(another Paul Burch idea, I think!) This was all great but after we finally matched a moving jeep, a moving airplane and a moving pickup, we forgot about one thing——the wind!

"To make a long story much shorter, the pickup came right up under the bird and we were caught in the wind breaking over the cab, due to a head wind we were fighting. The airplane was swept over the cab and down into the bed, flying, (I swear) less than two feet behind the back window! The guys in the pickup bed (the refueling crew) were all lying on their backs trying to become as small as possible, and all the while my prop turning at about 12,000 RPM was inches over their heads! At that moment I became the world's first gray-haired kid.

"Dad hollered at me that I was about to kill our fuel crew, as if I needed a reminder, and I pulled full back pressure on the control handle. As we broke out and climbed off above the pickup, one of the crew grabbed the plastic tube and pulled us backward! This created a second set of gray hairs. After a great tug-of-war we had fuel and were free again. I was happy to let Dad fly for a long time afterward.

"Dad handed off to me again and shortly after that I heard the engine stumble for the first time. Russ tightened up the circle he was driving to keep tension on the control lines and Dad told me to stay with it. It didn't take a genius to know that our flight was in big trouble. Our faithful Anderson Spitfire engine rapidly lost power and then simply quit. With a lump in my throat the size of a basketball I let her glide to slow down and then made one of the best landings of my life— then and now. With mixed emotions we all shook hands and the rest is history."   — BILL HART

The Yuma Daily Sun featured articles with accompanying photographs in advance of the flight of *The City of Yuma, Jr.* in its attempt to establish a world's endurance flight record for model airplanes. In the October 10, 1952 issue, *The Sun* article showed a headline: "Model Plane Stays Up 1 Hour, 10 Mins." It explains: "Ignition trouble brought the Jaycees' model airplane, *The City of Yuma, Jr.* down after one hour, ten minutes and 47 seconds, but the Jaycees were claiming the world's record in spite of the short run.

The article stated that Herschel Wright, driving the jeep, was able to establish contact with the model for four minutes while the crew of Odell Stafford and Harry Dye managed to get some gasoline into the airplane. Official timers for the model airplane flight were Dr. P. A. Birdick and "watch owners" Glenn Gartland and Gene Kornfeld, as *The Yuma Daily Sun* referred to two prominent jewelry store proprietors of that era. The airplane had a three-quart fuel capacity and the crew had hoped to keep it in flight for five hours, but oil got on the points of the motor and forced the flight to end. *The City of Yuma, Jr.* hung in the museum at the Yuma Territorial Prison for a number of years, but the prison was remodeled and the whereabouts of the model have been unknown since that time.

Ten years after their endurance flight, Bob and Woody were participants on *"I've Got a Secret,"* Garry Moore's popular national television show. They flew to New York for the show, which was arranged by the indomitable Ray Smucker, and their secret was not guessed by the panel of contestants. As a result of that show, the Sahara Hotel in Las Vegas sponsored a flight which broke the Yuma record.

Then, on April 27, 1991, the two pilots, Robert Woodhouse and Woodrow Jongeward, were inducted into *The Arizona Aviation Hall of Fame* at the Pima Air Museum near Tucson, at its second annual induction ceremony. The only four prior honorees were Frank Luke, Jr., Walter Douglas, Jr., Barry Goldwater, and Frank Borman. The endurance fliers' plaque states, in part: "A support team of 600 people was required for this highly-coordinated enterprise which was a punishing test of both men and machine."

In the future, scholars will look back at Yuma's progress through the 20th Century and they will note that the most significant of many turning points in Yuma's march to success occurred in 1949 with *"The Longest Flight."*

The manner in which the people of Yuma pulled together to make this happen and take Yuma forward was a high point that may never be equaled again. One fact stands out: Yuma, in 1949, was at the crossroads of mediocrity and success, and through the efforts of those six hundred volunteers, the path to success was made available. To honor those of the past that made this possible, the *"City of Yuma"* story was recently selected to be Yuma's entry to the White House Millennium Program, *"Honor the Past—Imagine the Future."* Judy Spencer provided the required information to Mike Shelton, Assistant

City Administrator, who, in turn, coupled this information with a proclamation from Mayor Young and sent it to Washington, D.C. These items will become a part of the White House Project Web Site, making *"The City of Yuma, Longest Flight"* known to the world once again. *"The City of Yuma"* will live forever.

— END —

Civil Air Patrol Cadets, Squadron 509

Travis Anderson	James Apple	Ben Boddy
William Bohi	Valentin Cortez	Corey Cowan
Jeff Dahle	Dustin Dinwiddie	Kimberly Dobbs
Daiv McBride	Omar Navarrete	Josh Smith
Robert Taylor	David Thiessen	Aileen Thompson